SpringerBriefs in Fire

Series Editor

James A. Milke

For further volumes:
http://www.springer.com/series/10476

Robert Upson · Kathy A. Notarianni

Quantitative Evaluation of Fire and EMS Mobilization Times

 Springer

Robert Upson
Department of Fire Protection
Worcester Polytechnic Institute
Worcester, MA 01609
USA

Kathy A. Notarianni
Department of Fire Protection
Worcester Polytechnic Institute
Worcester, MA 01609
USA

ISSN 2193-6595   ISSN 2193-6609   (electronic)
ISBN 978-1-4614-4441-1   ISBN 978-1-4614-4442-8   (eBook)
DOI 10.1007/978-1-4614-4442-8
Springer New York Heidelberg Dordrecht London

Library of Congress Control Number: 2012940266
Reprinted in 2012 by Springer Science+Business Media New York

Foreword

The ability of emergency response agencies to get personnel and equipment to the scene of an emergency in a timely manner is critical. This involves effective *alarm handling time* and *turnout time*. However, comprehensive data on emergency first responder alarm handling and turnout time is largely absent from the published literature.

Alarm handling time and turnout time are specific measurable segments of the overall *mobilization time* of emergency response units (along with *initiation time* and *travel time*). This study focuses on mobilization times involving alarm handling and turnout, i.e., the measureable time interval from call receipt at a public safety answering point until the first assigned emergency response unit is physically en route to the emergency.

Operational benchmarks for alarm handling and resource turnout would be greatly enhanced with strong empirical validation, and this information is of direct interest to the following four NFPA standards that address certain aspects of this topic: NFPA 450, *Emergency Medical Services and Systems*; NFPA 1221, *Public Fire Service Communications Systems*; NFPA 1710, *Career Fire Department Deployment*; and NFPA 1720, *Volunteer Fire Department Deployment*. It is generally understood that certain factors (e.g., notification methods, facility layout, tasks at time of alarm, etc.) will cause mobilization times to increase or decrease, but the importance and influence of these factors is not well known.

This study provides a quantitative evaluation of fire emergency and EMS mobilization times, and identifies key factors affecting their performance. It provides a statistical analysis of actual fire emergency and EMS alarm handling and turnout times based on data collected across a diverse representative population of North American fire service organizations. The results provide measured data for validation and refinement of requirements provided by nationally recognized standards, and additionally indicates the most significant and variable factors (e.g., difference in daytime and nighttime events).

The Research Foundation expresses gratitude to the report authors with the Department of Fire Protection Engineering at Worcester Polytechnic Institute (WPI). In addition, the Research Foundation appreciates the guidance provided by the Project Technical Panelists and all others that contributed to this research effort, especially the emergency first-responder organizations that participated in the data collection efforts. Special thanks to National Fire Protection Association (NFPA) for providing the funding for this project.

The content, opinions, and conclusions contained in this report are solely those of the authors.

Project Technical Panel

Paul Brooks, Center for Public Safety Excellence
William Bryson, Miami Fire-Rescue Ret. (NFPA 1710 Chair and Metro Chiefs)
Ken Burdette, Central Kitsap Fire and Rescue (NFPA 1221 IAFC Rep)
Frank Florence, NFPA (NFPA 450 Staff Liaison)
Robert Kilpeck, Brandon Fire Dept (NFPA 1720 NVFC Rep)
Ken Knipper, Melbourne, KY (NFPA 450 Chair)
Peter McMahon, Town of Grand Island (NFPA 1720 Chair)
Lori Moore-Merrell, IAFF
Steve Sawyer, NFPA (NFPA 1710/1720 Staff Liaison)
Bill Stewart, Toronto Fire Services (CAFC, IFE and Metro Chiefs)
Larry Stewart, NFPA (NFPA 1221 Staff Liaison)
Catherine Spain, National League of Cities (NFPA 1710 NLC Rep)
Stephen Verbil, Connecticut Dept of Public Safety (NFPA 1221 Chair)
Tom Wieczorek, ICMA (NFPA 1710 ICMA Rep)

Project Sponsor

National Fire Protection Association

Preface

Comprehensive data on fire emergency and EMS call processing and turnout time is largely absent from the published literature. Operational benchmarks for *alarm handling time* and *turnout time* specified in the NFPA peer consensus standards 1221 and 1710, respectively, would be greatly enhanced with strong empirical validation. This study presents a clear statistical picture of actual recorded *alarm handling times* and *turnout times* for fire and EMS emergencies across a group of large fire departments. Additionally, the study identifies some significant factors that affect variation in *alarm handling times* and *turnout times* in those departments. These results provide an objective basis for further development of the relevant codes and standards as well as contributing critical information for fire chiefs and other government decision makers tasked with optimum deployment of emergency response facilities (ERFs) and emergency response units (ERUs).

(i) The actual recorded *alarm handling times*, provided to this study from a group of large fire departments, were compiled, statistically analyzed, and compared to the target alarm handling times given in NFPA 1221. Results demonstrated that:

- For both fire and EMS calls, the mean average alarm handling times observed fell well within the current 60 s benchmark.
- For approximately 80 % of the fire and EMS calls, alarm handling was completed in the required 60 s or less.
- Eighty percent of calls processed in 60 s or less fall below the 90 % targeted in the standard.
- The time required for alarm handling of 90 % of the calls was 92 s for fire (slightly over one and one-half times the standard) and 84 s for EMS (slightly less than one and one-half times the standard).
- A second benchmark, which targets 90 s to process 99 % of the calls, is set in the standard. At an elapsed time of 90 s, approximately 90 % of the calls were processed rather than the 99 % required. Given the observed distribution of alarm handling times, where a very long tail is observed, the 99 % criterion

may not be particularly useful for benchmarking. A long tail is observed in the distribution, representing long alarm handling times for a certain fraction of the fire and EMS calls.

(ii) The actual recorded *turnout times*, provided to this study from a group of large fire departments, were compiled, statistically analyzed, and compared to the target alarm handling times given in NFPA 1710.

• For both fire and EMS calls, the mean average turnout times observed fell well within their respective current benchmarks; 80 s for fire and 60 s for EMS.

– For approximately 60 % of the fire calls, turnout was completed in the required 80 s or less.
– For approximately 54 % of the EMS calls, turnout was completed in the required 60 s or less.

• The time actually required and recorded for turnout of 90 % of the calls was 123 s for fire (slightly over one and one-third times the standard) and 109 s for EMS (slightly more than one and two-thirds times the standard).

(iii) The actual recorded *turnout times*, provided to this study from a group of large fire departments, showed a highly significant difference in *turnout times* between daytime and nighttime hours, a factor not currently addressed in NFPA 1710.

• Turnout times were compared between daytime hours (0600–1800), when crews are presumably at their highest readiness; and nighttime hours (0000–0600), when they are presumably at their lowest readiness.
• For both fire and EMS nighttime calls, the mean average turnout times observed fell well above their current NFPA 1710 benchmarks.

– For only approximately 21 % of the nighttime fire calls, turnout was completed in the required 80 s or less.
– For only approximately 12 % of the nighttime EMS calls, turnout was completed in the required 60 s or less.

• The time required for turnout of 90 % of the nighttime calls was 158 s for fire (just under two times the standard) and 144 s for EMS (slightly more than two-and one-third times the standard).

(iv) The simulated *turnout times* recorded in the Baseline Turnout Exercise, reported from a diverse group of fire departments, exceeded the benchmarks set in NFPA 1710.

• For simulated fire EMS calls, the mean average turnout times observed fell well within their respective current benchmarks: 80 s.

– For approximately 80 % of the exercise trials using the "wheels rolling" criterion, turnout was completed in the required 80 s or less.

- For approximately 70 % of the exercise trials using the "crosses sill" criterion, turnout was completed in the required 80 s or less.

- Both percentages of simulated turnouts completed in 80 s or less fall well below the 90 % targeted in the standard.
- The time actually required and recorded for turnout of 90 % of the calls was 86 s for the "wheels rolling" criterion and 96 s for the "crosses sill" criterion.

(v) The Station Layout Data collected indicates that the average station requires as much as twice the travel distance and time to reach the ERU from common station areas as is provided in the Baseline Turnout Exercise.

- Foot travel distance and time to sleeping areas is, on the average, significantly greater than travel distance to any other part of the ERF.
- Foot travel requires 10 s for every 50 feet traveled within the ERF, and stairs more than double that rate.

Acknowledgments

This study was funded by

Participating Fire Departments

Bainbridge Island Fire Department
  Bainbridge Island, WA
	Lexington Fire Department
	  Lexington, KY

Cary Fire Department
  Cary, NC
	Lincoln Fire and Rescue
	  Lincoln, NE

Chesapeake Fire Department
  Chesapeake, VA
	Orange County Fire and Rescue
	  Orange County, FL

Fairfax County Fire and Rescue Department
  Fairfax County, VA
	Southington Fire Department
	  Southington, CT

Flagstaff Fire Department
  Flagstaff, AZ
	Thornton Fire Department
	  Thornton, CO

Fort Worth Fire Department
  Fort Worth, TX
	Toronto Fire Services
	  Toronto, ON

Green Bay Fire Department
  Green Bay, WI
	Woodland Fire Department
	  Woodland, CA

Figures

Tables

Contents

Chapter 1
Introduction

A critical factor in the effectiveness of any emergency response agency is the ability to get personnel and equipment to the scene of the emergency in a timely manner. This response time can be roughly divided into two broad components: *mobilization time* and *travel time*. The current edition of NFPA 1710, *Standard for the Organization and Deployment of Fire Suppression Operations, Emergency Medical Operations, and Special Operations to the Public by Career Fire Departments* (NFPA 1710 2009) references three distinct time segments from NFPA 1221, *Standard for the Installation, Maintenance, and Use of Emergency Services Communications Systems* (NFPA 1221 2009). These segments are *alarm transfer time*, *alarm answering time*, and *alarm processing time*. Collectively those segments comprise *alarm handling time*.[1] NFPA 1710 further defines a segment referred to as *turnout time*. For the purposes of this study, these four segments together, measuring the time from call receipt at a *public safety answering point* (PSAP) until the first assigned ERU is physically en route to the emergency, will be referred to as *mobilization time*. The other segments identified in NFPA 1710, *travel time* and *initiating action/intervention time*, are outside the scope of this study.

NFPA 1221 defines a specific benchmark time for PSAPs and fire/EMS *communications centers* to process calls for emergency assistance. The current edition, NFPA 1221-2010, requires 90 % of calls to be processed within 60 s and 99 % of calls to be processed within 90 s.[2] NFPA 1710 defines a specific benchmark time for career fire departments to place their first *emergency response unit* (ERU) en route to an emergency. The current edition, NFPA 1710-2010, requires the first EMS ERU to be en route within 60 s 90 % of the time and the first fire ERU to be

---

[1] *Alarm Handling Time* was identified as *Call Processing Time* in the edition of the NFPA 1221 standard in effect when this study was begun. The use of *call processing time* in any documents related to this study should be viewed as functionally interchangeable with the newer terminology, *alarm handling time*.

[2] NFPA 1221-2010: 7.4.2* 90 % of emergency alarm processing shall be completed within 60 s, and 99 % of alarm processing shall be completed within 90 s. (*For documentation requirements, see* 12.5.2.)

R. Upson and K. A. Notarianni, *Quantitative Evaluation of Fire and EMS Mobilization Times*, SpringerBriefs in Fire, DOI: 10.1007/978-1-4614-4442-8_1,
© Fire Protection Research Foundation 2010

en route within 80 s 90 % of the time measured from the beginning of alert notification.[3]

To a large extent these benchmark times are based on qualitative data, experience, and assumptions and do not have a strong body of empirical data to justify them. Preliminary data[4,5,6] shows that these times may be unrealistically short in today's fire service environment and may lead to errors in analyses used to determine future station locations and determine mobile resource allocations; discourage fire departments from trying to meet the performance objectives in these NFPA standards; and encourage unsafe practices in an effort to meet unrealistic alarm handling and turnout objectives.

What is Mobilization Time?

Mobilization Time combines two related response time segments identified by separate NFPA standards: *Alarm Handling Time* (formerly identified as Call Processing Time) and . *Mobilization Time* represents the period of time beginning when a call for emergency aid is received and ending when appropriate fire department emergency response units (ERUs) are actually en route to the emergency. The component parts of this study's *Mobilization Time* construct as they are defined by NFPA 1221 and NFPA 1710 are as follows:

Alarm Handling Time: The time interval from the receipt of the alarm at the primary PSAP until the beginning of the transmittal of the response information via voice or electronic means to emergency response facilities (ERFs) or the emergency response units (ERUs) in the field.

---

[3] NFPA 1710-2010:
    4.1.2.1 The fire department shall establish the following objectives:

(1) Alarm handling time to be completed in accordance with 4.1.2.3.
(2) 80 s for turnout time for fire and special operations response and 60 s turnout time for EMS response.

    4.1.2.4 The fire department shall establish a performance objective of not less than 90 % for the achievement of each turnout time and travel time objective specified in 4.1.2.1.

[4] In 38 simulated turnout exercise trials conducted with career fire units in conjunction with the "Multi-Phase Study on Firefighter Safety and Deployment" (Averill et al. 2008), Upson found a mean turnout time of 70 s with only 80 % of the simulated turnouts at or below the 80 s specified by NFPA 1710 (Upson 2009).

[5] An investigation of turnout time in 15 career fire departments found a mean turnout time of 81 s for fire responses and 69 s for EMS responses between 0700 and 2200, with only 34 % of fire response within 60 s and only 45 % of EMS responses within 60 s (CPSE/Deccan International 2007).

[6] In a study of 100 observed responses over a 3-month period, Gill reports only 85 % of turnout times under 81 s (Gill, IRMP Year III—Turnout Times 2007) (Gill 2009).

- *Alarm Transfer Time*: The time interval from the receipt of the emergency alarm at the public service answering point (PSAP) until the alarm is first received at the communication center.
- *Alarm Answering Time*: The time interval that begins when the alarm is received at the communication center and ends when the alarm is acknowledged at the communication center.
- *Alarm Processing Time*: The time interval from when the alarm is acknowledged at the communication center until response information begins to be transmitted via voice or electronic means to emergency response facilities (ERFs) and emergency response units. (ERUs)

Turnout Time: The time interval that begins when the emergency response facilities (ERFs) and emergency response units (ERUs) notification process begins by either an audible alarm or visual annunciation or both and ends at the beginning point of travel time. (Travel Time: The time interval that begins when a unit is en route to the emergency...)

(NFPA 1710 2009)

During this study, researchers collected, organized, and analyzed large quantities of recorded historical alarm handling time and turnout time data. The results of those analyses can be used by code and standards development committees, fire chiefs, and other government decision makers to provide an objective basis for establishing benchmark times in consensus standards; for efficiently designing and locating fire stations; and for other appropriate allocations of resources.

Chapter 2
Study Questions and Research Methods

The questions addressed in this study are presented below along with the research methods employed in an effort to answer them. The study questions fall into two categories:

- Time-to-Task Completion: How long do certain tasks typically take to perform?
- Factors Influencing Mobilization Time: What other factors influence time-to-task performance?

2.1 Time-to-Task Completion

There are six primary mobilization time-to-task questions addressed in this study:

I. *In a representative group of career or mostly career fire departments, what is the time actually spent completing alarm handling?*
II. *How does actual recorded alarm handling data compare to the NFPA 1221 standard benchmarks?*
III. *In a representative group of career or mostly career fire departments, what is the actual time typically required for turnout?*
IV. *How does the actual recorded turnout time data compare to the ŃFPA 1710 standard benchmarks for turnout time?*
V. *In a representative group of career or mostly career fire departments, what is the actual time typically required for mobilization?*
VI. *How does the actual recorded turnout time data compare to an implied hypothetical NFPA standard benchmark for mobilization time?*

R. Upson and K. A. Notarianni, *Quantitative Evaluation of Fire and EMS*
Mobilization Times, SpringerBriefs in Fire, DOI: 10.1007/978-1-4614-4442-8_2,
© Fire Protection Research Foundation 2010

2.2 Factors Influencing Mobilization Time

In the standards making process, it is important to understand both the processes and the factors that influence them when functional performance objectives are established. Alarm handling and turnout have traditionally been addressed as separate processes with separate time objectives, but they are functionally connected by a critical communications link. Emergency call takers and dispatchers tasked with alarm handling must collect sufficient and accurate information to identify and alert the appropriate emergency response units (ERUs) for each call for emergency aid and then communicate that information in a clear, timely manner. The ERUs must receive and accurately interpret that information quickly in order to turnout efficiently without delays introduced by miscommunications or missed communications. ERUs must operate from an emergency response facility (ERF) designed to facilitate the receipt of those crucial communications and not hinder efficient turnout.

This study examined six specific factors influencing Mobilization Time:

- Combined PSAP/Communications Center Versus Separate Locations/Agencies

  - Does the introduction of a transfer between emergency call takers and dispatchers increase *alarm handling time*?

- Voice-Only Dispatch Versus Dispatch to Printer or Mobile Display Terminal (MDT)

  - Does the presence of clear written dispatch information improve *turnout time*?

- Fire Response Versus EMS Response

  - Does preparing for a fire response require a longer *turnout time* than for an EMS response?

- Daytime Versus Nighttime Response

  - Do nighttime turnouts require significantly more time than daytime turnouts?

- Firefighter Crew Proficiency in Baseline Turnout Exercise

  - Have we accurately assessed the time needed for turnout under ideal conditions?

- Effects of Station Layout on Turnout Response

  - How much does the size and layout of ERFs affect *turnout time*?

2.3 Research Methods

This study was able to share resources with the ongoing Department of Home land Security funded "Multi-Phase Study on Firefighter Safety and Deployment of Resources" to develop a representative pool of prospective participants for the collection of historical response data, baseline turnout exercise trials, and station information. The more than 400 agencies represented in that study were randomly selected using a statistical model to represent fire departments of various size throughout the United States, and a large pool of fire department demographic information was already being collected and was made available to this study.

A subset of the Firefighter Safety and Deployment sample population was asked to participate in this study based on the availability of communications center data documenting all four time segments making up the mobilization interval. Maintenance of that documentation is a requirement of the NFPA 1221 standard, and a significant number of suitable sample departments were expected to participate. In addition to gathering historical response data on mobilization, a survey was created to identify the effects of six specific factors potentially affecting variance in mobilization time.

- Combined PSAP/Communications Center Versus Separate Locations/Agencies
- Voice-Only Dispatch Versus Dispatch to Printer or MDT
- Fire Versus EMS Response
- Daytime Versus Nighttime Response
- Firefighter Crew Proficiency in Baseline Turnout Exercise
- Potential Effects of Station Layout on Turnout Response

The Mobilization Study was accomplished in five steps:

1. Selection of a representative cross-section of participant departments from the Firefighter Safety and Deployment sample sufficient to include at least 100 ERFs
2. Develop survey form in consultation with project technical advisors

    2.1. Information on PSAPs and communications centers
    2.2. Dispatch methods used/pertinent standard operating procedures

3. Data Collection

    3.1. Collection of survey form data
    3.2. Collection of one year of historical response data in electronic format
    3.3. Collection of Baseline Turnout Exercise data by FD representatives
    3.4. Collection of Station Information by FD representatives
    3.5. Field visits to selected departments to spot check documentation methods

4. Analysis of Historical Response Data

    4.1. Means, Standard Deviations, Cumulative Data Function Plots
    4.2. Alarm handling times by type of call and time of day
    4.3. Turnout times by type of call and time of day

        4.3.1. Identify factors creating variance and relative significance of each factor
        4.3.2. Correlate mobilization times with factors from survey data

4.3.3. Compare actual turnout time data with results from Baseline Turnout Exercise

5. Preparation of Reports

5.1. Final Report to Fire Protection Research Foundation
5.2. Summary Reports to NFPA 1221 and NFPA 1710 Technical Committees

Chapter 3
Recruiting Participants

3.1 Invitations

There are over 4,000 career or mostly career fire departments in the United States. Representing only 14 % of U.S. fire departments, these departments nonetheless protect 61 % of the population (U.S. Fire Administration 2008). The "Multi-Phase Study on Firefighter Safety and Deployment of Resources (Averill et al. 2008)" study sample was randomly selected to represent a statistically balanced cross section of career and mostly career U.S. fire departments[1] and included departments that had already indicated a willingness to participate in similar fire-service research projects. We extended invitations to all 457 fire departments selected into that sample with letters sent by U.S. Postal Service addressed to the chief officer of record of each department.

---

[1] "The sample of departments was determined in two steps—(a) identification of 'selfrepresenting' departments, and (b) the selection of the remaining 'non-selfrepresenting' departments. Self-representing departments are those whose populations served (and thus annual fire/EMS event volume) is so large that they would appear in the sample with certainty. To understand why self-representing departments are unavoidable in our probability-proportional-to-size (pps) sample, note that a sample of 494 departments would cover in aggregate the 227 million population served. Accordingly, each sampled department will 'represent' (227 million)/494 or about 460,000 population. Thus any department whose population served exceeds 460,000 would be sampled with certainty. It is conventional in survey sampling to use 75 % of this ratio as a threshold for determining self-representing selections. We used this approach to identify 76 self-representing departments, all with 'population served' exceeding 350,000.

"The remaining 418 non-self-representing departments were stratified by population served and sampled with probabilities proportional to population served. Combined with the 79 self-representing departments, this yielded our total desired sample size of 494." (Averill, et al. 2008, 57)

R. Upson and K. A. Notarianni, *Quantitative Evaluation of Fire and EMS Mobilization Times*, SpringerBriefs in Fire, DOI: 10.1007/978-1-4614-4442-8_3,
© Fire Protection Research Foundation 2010

3.2 Screening Questionnaire

In addition to recruiting and verification of contact information, the initial participant screening questionnaire collected basic data on two areas: basic call intake methods and availability of computer readable alarm handling documentation.

Prospective participants were asked to identify how 9-1-1, direct dialed, and automatic alarms were initially received by their departments. They indicated if the fire department, police department, or another agency was responsible for handling the intake of emergency calls for aid for their department.

Prospective participants were also asked to indicate whether they documented alarm handling times in accordance with NFPA 1221 and whether they were available in an exportable computer format.

Of the 79 respondents to the Screening Questionnaire in the affirmative, 59 also responded to an email confirmation sent to their designated contact person. Those 59 departments advanced to the next phase of the study.

3.3 Participant Survey

The Participant Survey was used to collect demographic information about each department to establish the range of diversity within the potential study sample and to identify departments collecting well documented historical response data suitable for use in this study.

The 59 departments' designated contacts were supplied with online usernames and passwords to access the web-based data collection site at Worcester Polytechnic Institute (WPI). Login instructions and a written list of Participant Survey questions were provided by email. Participants were asked to complete the Participant Survey online.

The survey included 43 questions in five categories:

- Fire Department Demographics—basic descriptive information including questions about fire and EMS call volume.
- Alarm Handling and Emergency Dispatching System—information regarding the 9-1-1 call intake agencies, direct dialed calls, and automatic alarms; calls dispatched by answering agency or transferred to a separate Communications Center; and dispatcher training.
- Alarm Handling Documentation—information regarding time documentation; and manual and automatic timestamping of records.
- Emergency Dispatching Documentation—methods used to notify emergency units of alarms; and manual and automatic timestamping of records.
- Emergency Response Unit Notification—methods used to receive and record emergency notifications in station; and paging, print-outs, and MDTs.

Of the 59 participant departments, 38 accessed the survey online and completed at least a portion of the survey questions. Eight departments of those 38 were eliminated due to incomplete survey responses. Of the remaining 30 departments, 20 were selected to advance to the main phase of the study based on their responses to the Participant Survey.

Of the final 20, 12 departments responded that they documented all of the key response information ideally required for participation in this study:

- Recorded call priority designation, specifically responses designated as "emergencies."
- Recorded and time-stamped receipt of emergency calls for aid at the point when the call was answered.
- Recorded and time-stamped dispatch of emergency responses at the point when the dispatch data was transmitted to the ERFs/ERUs.
- Recorded and time-stamped "en route" notifications from emergency response units.

Most of the ideal candidate departments directly managed all their alarm processing from answering the initial call for emergency aid through en route notification by ERUs. In order to include more participants that split alarm handling between separate PSAPs and Communication Centers, eight additional departments using separate PSAPs and Communications Centers were invited to provide comparison data. These departments typically had less accurate documentation methods for en route times but were otherwise good sources for comparative data.

3.4 Final Participants

Of the 20 departments invited to participate in the main phase of the study, 14 departments returned data from at least one of the requested areas (Historical Response Data, Baseline Turnout Exercise, and Station Information) in time to be included in the study analyses. These departments were geographically diverse (Fig. 3.1 Geographic Distribution of Final Participants) and adequately represented departments of varied size (Fig. 3.2 IAFC Department Class). The final list of departments served a range of populations ranging from 23,000 to 2.5 million (median, 221,000) and operated from 1 to 84 career stations (median 14).

This sample provided ample data to examine four of the six factors targeted for study:

- Fire Versus EMS Response.
- Daytime Versus Nighttime Response.
- Firefighter Crew Proficiency in Baseline Turnout Exercise.
- Potential Effects of Station Layout on Turnout Response.

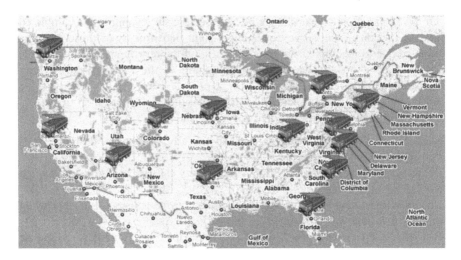

Fig. 3.1 Geographic Distribution of Final Participants. Bainbridge Island Fire Department, Bainbridge Island, WA. Cary Fire Department, Cary, NC. Chesapeake Fire Department, Chesapeake, VA. Fairfax County Fire and Rescue Department, Fairfax County, VA. Flagstaff Fire Department, Flagstaff, AZ. Fort Worth Fire Department, Fort Worth, TX. Green Bay Fire Department, Green Bay, WI. Lexington Fire Department, Lexington, KY. Lincoln Fire and Rescue, Lincoln, NE. Orange County Fire and Rescue, Orange County, FL. Southington Fire Department, Southington, CT. Thornton Fire Department, Thornton, CO. Toronto Fire Services, Toronto, ON. Woodland Fire Department, Woodland, CA

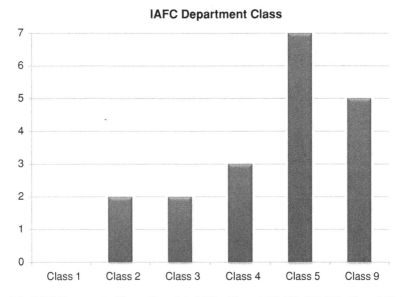

Fig. 3.2 IAFC Department Class. *Class 1* 0–9,999. *Class 4* 100,000–199,999. *Class 2* 10,000–49,999. *Class 5* 200,000 and up. *Class 3* 50,000–99,999. *Class 9* eligible for Metropolitan Chiefs

There was insufficient data to reasonably compare the other two of the originally targeted factors:

- Combined PSAP/Communications Center Versus Separate Locations/Agencies

  - None of the participant departments utilizing separate agencies for PSAPs and Communications Centers (i.e., those transferring a call from an initial PSAP Emergency Call Taker to a separate agency Communications Center Dispatcher), were able to provide time documentation from the PSAP, making a complete assessment of alarm handling impossible.

- Voice-Only Dispatch Versus Dispatch to Printer or MDT.
- Most departments in the final sample utilized apparatus mounted MDTs as well as voice dispatch. There was insufficient representation of departments utilizing voice-only dispatching to make a valid comparison.

Finally, as a quality-assurance measure, visits were made to two participating departments to verify the methods used to document *alarm handling* and *turnout times* and observe firsthand how typical fire and EMS responses were handled in those departments. The Fort Worth, TX, and Flagstaff, AZ, fire departments graciously consented to host visits to their departments, including a visit to Fort Worth's fire-dispatch center to better understand their advanced alarm handling system; touring several fire stations in both cities; interviewing line officers about turnout and en route notifications; and witnessing typical emergency responses. These visits provided valuable insight into the dynamics of alarm handling and turnout styles in those two participating departments.

Chapter 4
Data Collection

4.1 Historical Response Data

Participant departments were asked to provide historical response records of emergency incidents categorized as *Fire*, *EMS*, or *Other*. The determination of what type of call constituted an "emergency response" was left the participant department consistent with the definition in NFPA 1710.[1] Specifically, departments were asked to provide computer data files that included the following:

- Only data from responses designated as emergency priority.
- Responses identified as *Fire*, *EMS*, or *Other* types of emergency.
- Time of call receipt (critical: begins the *alarm handling time* interval).
- Time of call transfer from PSAP to Communications Center (if applicable, delimits *alarm transfer time* interval).
- Time of dispatch "Pre-Alert" (if applicable and available).
- Time of dispatch (critical: ends the *alarm handling time* and begins the *turnout time* interval).
- Identification of "First Due" apparatus assigned.
- Time of "First Due" apparatus acknowledgement (if available).
- Time of "First Due" apparatus en route (critical: ends the *turnout time* interval).
- NFIRS incident number for the response (if available).

Twenty departments were asked to supply historical response records of emergency incidents categorized as Fire, EMS, or Other. Data were submitted by

---

[1] NFPA 1710–2010:

3.3.41.1 *Emergency Operations.* Activities of the fire department relating to rescue, fire suppression, emergency medical care, and special operations, including response to the scene of the incident and all functions performed at the scene [1500, 2007] (NFPA 1710 2009).

R. Upson and K. A. Notarianni, *Quantitative Evaluation of Fire and EMS Mobilization Times*, SpringerBriefs in Fire, DOI: 10.1007/978-1-4614-4442-8_4, © Fire Protection Research Foundation 2010

Table 4.1 Summary of data collected

Participating department (population served)	Response records collected[a]	Station data collected[b]	Turnout exercises collected[c]	Emergency call intake PSAP[d]	Alarm handling method[e]	EMS transport[f]
Bainbridge Island, WA (23,000)		1	2			
Cary, NC (135,000)		1	9			
Chesapeake, VA (221,000)	57,552	15	10	Police department	Transfer	Yes
Fairfax County, VA (1.1 million)	56,219	37 (of 50)	3	Fire department	Direct	Yes
Flagstaff, AZ (65,000)	9,898	6 (of 7)	6	Police department	Direct	No
Fort Worth, TX (270,000)	86,299	20 (of 41)		Police department	Transfer	No
Green Bay, WI (104,000)		7	14			
Lexington, KY (261,000)	34,811	16 (of 22)	2	Local PSAP	Transfer	Yes
Lincoln, NE (250,000)	13,903	2 (of 14)	9	Local PSAP	Direct	Yes
Orange County, FL (856,000)	78,506	29 (of 41)	30	Fire department	Direct	Yes
Southington, CT (43,000)		1	4			
Thornton, CO (118,000)	8,105	5	16	Fire department	Transfer	Yes
Toronto, ON (2.5 million)	183,247	82	12	Police department	Transfer	No
Woodland, CA (53,000)	4,835	3	8	Regional PSAP	Direct	No
14	533,376	225	125	3 FD, 4 PD, and 3 (non-FD/PD) PSAPs	5 Direct 5 Transfer	6 Yes 4 No

[a] Number of raw electronic response records submitted by participating department
[b] Number of Station Information Sheets (one per station) submitted by participating department
[c] Number of Baseline Turnout Exercise trials submitted by participating department
[d] Primary agency responsible for receiving 9-1-1 calls for participating department (Fire, Police, non-FD/PD PSAP agency)
[e] PSAP agency dispatches fire calls directly or transfers call to separate agency Communications Center
[f] Participating department provided EMS transport (ambulance) services (Yes or No)

the departments using a variety of media, including email, CD, and thumbdrive, in .xls or .csv formatted data.

Half of the participating departments provided a total of 533,376 response records for analysis. These responses represented 1 year of emergency responses from each of these departments (Table 4.1).

Four of the 20 participating departments were ultimately unable to provide response records within the time available.

4.2 Baseline Turnout Exercise

Participating departments were provided with instructions to layout and conduct a standardized Baseline Turnout Exercise. They were provided with a stopwatch to accurately measure the Baseline Turnout Exercises along with standard data collection sheets to record the results. Finally, they were provided with a prepaid mailer to return the data collection sheets.

The Baseline Turnout Exercise consists of two parts:

- Part 1: *Alarm Response Walking Speed*. This is a timed exercise where fire crews in station uniforms walk over a measured course at the pace they had typically use when moving to their fire/EMS apparatus during turnout.

1. The crew moves "quickly and with purpose"; no jogging or running; "As you would normally respond to an emergency call in your station."
2. The crew starts on "ready, set, go" command.
3. The crew proceeds to the front end of the bay; crosses over the measured line return; returns and crosses over the starting line to finish.
4. Using a stopwatch, record times for the first and last crew member over the finish line. (If the entire crew crosses the line too closely to reliably time both first and last, record the first time only.)

- Part 2: *Scramble, Don, & Mount*. A timed exercise where fire crews in station uniforms simulate a complete turnout in a standardized layout from notification of the alarm, to donning PPE (Personal Protective Equipment), to mounting the apparatus, to physically leaving the station.

1. Starting configuration

- "Le Mans Start"—Crew starts shoulder to shoulder in a line facing away from the apparatus at the back of the bay
- Crew is dressed in regular station wear
- PPE is stowed at each crew member's riding position[2]
- Apparatus is parked with its front 5' from the line of the door sill

---

[2] EMS crews may omit PPE as per local SOP. Fire suppression crews must don full PPE including either a helmet or radio headset.

- MPO & Officer's windows are open
- Apparatus is otherwise as it would normally be stowed in station

2. On command ("Go!", no countdown preparation)

   - Timer must be positioned to observe "wheels rolling" safely before MPO can release brakes!

3. Special Tasks

   - Officer retrieves "Run Sheet" from simulated printer
   - MPO opens bay door (manually or with remote)

4. Crew moves promptly to gear; all crew members don core gear at minimum (see footnote 2)

   - Bunker pants
   - Boots
   - Flame retardant hood (if normally worn)
   - Bunker coat
   - Helmet (or headset if that is standard practice)

5. Crew mounts apparatus

   - MPO may start engine at any time

6. Crew will not don SCBA per consensus of technical advisors
7. Crew fastens seatbelts

   - Seatbelt compliance confirmed by apparatus officer
   - MPO shall not release brakes until compliance confirmed

8. MPO pulls straight forward promptly (Maximum Speed 10 mph)

   - Brakes released; wheels start (Split 1)
   - Front Bumper crosses doorway sill marked by traffic cones (Split 2)
   - MPO must stop before reaching curb line marked by traffic cones[3]

Twenty departments were asked to conduct and record as many Baseline Turnout Exercises as reasonably possible within the allotted timeframe for data collection.

Thirteen departments provided a total of 125 turnout exercise trials for analysis, representing data from over 300 career firefighters in the Alarm Response Walking Speed exercise and over 100 career fire and EMS crews of two to four members in the Scramble, Don, & Mount response exercise (Table 4.1). All data were submitted via postal mail or scanned and returned by email.

---

[3] Where the proximity of public sidewalks and streets limits the distance the apparatus can continue out of the bay, appropriate modifications should be made. Please note any such modifications on the data collection sheet.

4.3 Station Information

Participating departments were asked to provide information on each of their ERFs on the station information sheets developed for this study. The information collected included each station's identification, address, a characterization of the station's population served as urban, suburban, or rural, and an inventory of apparatus based at that station.

Instructions were also provided to measure and record key travel distances within each fire station. Participants were instructed to measure from the apparatus driver's door to the day room, training room, dining/kitchen area, fitness room, and most remote bunk in the sleeping area. A prepaid mailer was included to return the Station Information Sheets.

> What is the average typical travel distance from each listed area in the fire station to the driver's door of each type of first-due apparatus? If there is more than one first-due suppression apparatus or EMS unit, list the average distance by type. Distances should reflect a typical path of travel for emergency personnel responding from each area. A surveyor's wheel would be ideal for this measurement. Please make your best estimate.

Twenty departments were asked to complete a Station Information Sheet about all of their ERFs or as many as reasonably possible within the allotted timeframe for data collection. Fourteen departments provided layout and foot travel data for a total of 225 fire stations, which well exceeded the study target of at least 100 stations (Table 4.1). These data include a wide variety of station layouts and designs ranging from historical stations, traditional urban stations, modern suburban stations, temporary stations, and at least one dual purpose suburban/crash fire rescue airport station.

Chapter 5
Data Analysis

5.1 Data Preparation

Raw data files submitted by each participant department were imported into an Excel® (Microsoft Corporation 2006) workbook file and converted into a standardized record format combining data from the initial screening survey, participant questionnaire, and historical response data into a single spreadsheet. Basic calculations and categorizations were performed and added to the spreadsheet in separate columns:

- Each response was categorized as *Day* (0600–1800), *Evening* (1800–0000), or *Night* (0000–0600) based on the recorded alarm time.
- Each response was categorized as *Fire*, *EMS*, or *Other* based on the classification submitted. Some participants provided detailed NFIRS or similarly detailed response classifications, which were systematically reclassified into the corresponding broad *Fire*, *EMS*, or *Other* category.
- *Alarm handling times*, *turnout times*, and *mobilization times* were calculated and recorded in whole seconds for each response record from the alarm, dispatch, and en route times provided.
- Where responding apparatus identifications were provided and cross referenced data were available from the Station Information Sheets, ERUs in each response record were:

  - Categorized as Engine, Truck, Rescue, Ambulance, or Other.
  - Matched with their station of origin and matching station layout data.

Data supplied on Baseline Turnout Exercise and Station Information Sheets were entered into each department's Excel file. Individual department summaries were calculated for the Baseline Turnout Exercise and added to the combined data spreadsheet for each department. Station Layout data from the Station Information Sheets was arranged as a lookup table to match up each apparatus from the

R. Upson and K. A. Notarianni, *Quantitative Evaluation of Fire and EMS Mobilization Times*, SpringerBriefs in Fire, DOI: 10.1007/978-1-4614-4442-8_5,
© Fire Protection Research Foundation 2010

historical response records with its normally assigned station and associated station layout measurements, if available. Turnout Exercise data from all reporting departments were combined in a separate Excel workbook for aggregate analysis.

The combined data spreadsheet from each workbook file were exported as a text file in comma separated value (.csv) format for import into SAS.

5.2 SAS Procedures

The Excel-generated text files containing each department's standard format data were imported into SAS® (SAS Institute Inc. 2003) via a text file (.csv). Key study data for analysis—*department identification, call type, time of day, alarm handling time, turnout time,* and *mobilization time*—were extracted by Query and stored as individual SAS data files for each participating department. Data from departments matched by alarm handling method (*Direct* or *Transfer*) and response documentation methods was combined into master data sets for statistical analysis.[1]

"Combined Set A" (n = 153,463)[2] was created by combining data from departments collecting response data with similar methodologies during the mobilization process from receipt of call, to dispatch of emergency response units, to designation of en route status. These departments were matched based on the following criteria[3]:

- All the departments represented managed combined PSAP and Communication Centers as single agencies providing documentary continuity from *alarm time* to *en route* time.
- *Alarm time* was timestamped and recorded electronically at the time the call was answered.
- *Dispatch time* was timestamped and recorded electronically at the time dispatch was initiated for all but one department that timestamped as dispatch was completed.
- *En route time* was timestamped and recorded electronically at the time responding units manually signaled en route from apparatus mounted MDTs.

---

[1] The data analysis for portions of this paper was generated using SAS/STAT software, Version 9, of the SAS System for Windows. Copyright © 2003 SAS Institute Inc. SAS and all other SAS Institute Inc. product or service names are registered trademarks or trademarks of SAS Institute Inc., Cary, NC, USA.

[2] One department with qualifying call-handling and time-documenting methodology was omitted from this data set due to issues with corrupted data formatting in the raw data file supplied by the department. It may be possible to recover this data for future analyses, but it was not practical within the time constraints of this study.

[3] It was not known at the time of data collection but, upon inquiry during the analysis phase, it was determined that all of the departments whose data was used in "Combined Set A" also report that they regularly review alarm handling and turnout times against NFPA or similar benchmarks as performance goals.

This set contains response data from four departments with a combined population served of 2,259,000 (median 553,000) from 108 stations. 97 % of the response records in this set are from IAFC Class 5 fire departments.

"Combined Set B" (n = 370,014) was created by combining departments with PSAPs that transferred calls for aid to separate Communication Centers for dispatching. These departments were matched based on the following criteria:

- All the departments represented utilize separate and distinct PSAPs and Communication Centers. PSAPs handle calls from *alarm time* to *alarm transfer* and Communications Center handle calls from *transfer time* through *en route time*. None of the participating departments in this group was able to provide *alarm times* from their PSAPs; thus we were unable to calculate *alarm handling times* from this group.
- *Dispatch time* was timestamped and recorded electronically at the time of dispatch for all departments in this data set.
- *En route time* was timestamped and recorded electronically at the time responding units manually signaled en route from apparatus mounted MDTs for some departments and recorded manually based on radio transmissions from responding units for others. This difference limited aggregation of response data for analysis in this group.

This set contains response data from five departments with a combined population served of 3,370,000 (median 265,500) from 160 stations. Ninety-eight percent of the response records in this set are from IAFC Class 5 fire departments.

"Combined Set X" (n = 523,477) was created by merging Combined Sets A and B. This data set was used for some very preliminary turnout time analysis but was discarded due to the variability of time documentation methods among departments and the inability to follow whole responses through both alarm handling and turnout.

Tabular data shown for Fire Versus EMS and Daytime Versus Nighttime analyses was produced from SAS using PROC UNIVARIATE and PROC FREQ.

Bar graphs, CDF plots, and distribution plots shown for Fire Versus EMS and Daytime Versus Nighttime analyses were produced in Excel based on data generated in SAS using PROC FREQ. Continuous CDF and distribution plots were drawn from 5-s interval data points with plot lines interpolated by Excel using the "smoothed lines" option. Bar graphs were generated from data binned as indicated on the graphs.

CDF plots shown for the Baseline Turnout Exercise were produced in Excel and were drawn from 5-s interval data points with plot lines interpolated by Excel using the "smoothed lines" option.

5.3 Quality Analysis

In order to eliminate idiosyncratic artifacts from the study data set, responses where either the *alarm handling time* or *turnout time* were recorded as less than or equal to zero were eliminated from statistical analyses. It was the consensus of the

investigators and project technical panel that such calls represent unusual cir-
cumstances that are not representative of normal response times.

- All analyses shown in this document have been conducted with responses where
  both *alarm handling time*, t_{AH}, and *turnout time*, t_T, are greater than zero.
- $t_{AH} > 0$ *and* $t_T > 0$

In an attempt to identify and remove large values suspected of being artifacts,
analyses were conducted initially on both raw data and data filtered. There were
reservations among members of the project technical panel about systematically
omitting large values, and comparison of the raw versus filtered analyses did not
show significant differences in overall results. Based on those early results, it was
decided that no responses would be omitted based on large values of either *alarm
handling time* or *turnout time*. This is discussed in more detail in the Appendix B.

5.4 Assumptions and Limitations

The data collected by this study is representative only of the sample collected,
namely, a small group of geographically diverse career fire departments serving
populations ranging in size from 23,000 to 2.5 million. Although the original 457
departments invited to participate in this study represented a statistically robust
randomized sample of US departments, the final sample represents only a fraction
of that original sample profile. Ultimately, selection of the final sample was
strongly influenced by the departments' willingness to commit time and resources
to the study and the availability of useable historical response records.

Historical Response Records

The quality of the data provided in these records is limited to the accuracy and
reliability of the data recorded by the participating departments and provided to the
study authors.

- The documentary timestamps that create the endpoints for observed *alarm
  handling time*, *alarm time* and *dispatch time*, were computer generated in the
  participant sample, "Combined Set A," and, as such, are assumed to be reasonably reliable and valid representations of actual *alarm handling times*.
- There is a greater degree of uncertainty affecting the reliability and validity of
  the *en route time* timestamps on which observed *turnout time* is based. This
  uncertainty is due to both human and technical factors involved in transmitting
  the en route signal.

Firefighters/fire officers have commonly reported signaling en route via the MDT several seconds before the apparatus is actually in motion[4] and there is also some concern about transmission delays introduced by complicated communications systems in recording MDT generated timestamps.[5] This is a known limitation of the available data but has been assumed to be a reasonable approximation when averaged over a substantial number of responses.

- The data sets used in Sects. 6.1 and 6.2, are based on "Combined Set A" and, as such, are strongly representative of IAFC Class 5 fire departments. All of our findings share the limitations of the data collection sample; thus extrapolation of the data to predict mobilization times beyond the sample for smaller fire departments or for volunteer departments may not be valid.

Baseline Turnout Exercise

It is assumed that the exercise design represents a typical set of turnout tasks that can be reasonably assumed to apply to any fire department response.

All participants were provided with written instructions and a stop watch in order to maintain a reasonably uniform standard for the timed exercise across participating departments. Participants conducting the drills generally understood and followed the directions as intended, and reasonable care was taken to record accurate times.

Station Information

Participants providing station information, particularly with regard to measuring travel distances, understood and followed the directions as intended and took reasonable care to record accurate information.

---

[4] During informal interviews, one engine officer mentioned that his driver typically would reach over to the MDT and signal "en route" while the officer was donning his PPE. Others indicated that they would signal "en route" as soon as they were seated. Such early reports seem especially common in departments where turnout times are regularly reviewed against benchmark standards.

[5] One study found that the transmission processing delay introduced in an 800 MHz communications system delayed recording an MDT-generated en route signal by an average of 2 s with some delays ranging as high as 7 s (Office of Strategic Planning and Information Systems of the Greensboro (NC) Fire Department; Guilford Metro 911 Emergency Communications Center 2007).

5.5 Data Sets Applied to Study Questions

As presented in Sect. 2.1, there are six primary mobilization time questions addressed in this study:

I. In a representative group of career or mostly career fire departments, what is the time actually spent completing alarm handling?
II. How does actual recorded alarm handling data compare to the NFPA 1221 standard benchmarks?
III. In a representative group of career or mostly career fire departments, what is the actual time typically required for turnout?
IV. How does the actual recorded turnout time data compare to the NFPA 1710 standard benchmarks for turnout time?
V. In a representative group of career or mostly career fire departments, what is the actual time typically required for mobilization?
VI. How does the actual recorded turnout time data compare to an implied hypothetical NFPA standard benchmark for mobilization time?

In addition to addressing those questions, this study set out to examine specific factors influencing mobilization time:

• Fire Versus EMS Response
• Daytime Versus Nighttime Response
• Firefighter Crew Proficiency in Baseline Turnout Exercise
• Effects of Station Layout on Turnout Response

After a careful review of the data collected, the authors elected to address all six questions in pairs; looking at question pairs I and II, III and IV, and V and VI (as above) for fire and EMS separately. This method was intended to provide insight into the six main study questions and identify any differences in the type of call factor, namely, Fire Versus EMS Response.

5.5.1 Fire Versus EMS Response

This analysis was drawn from the Historical Response Data contained in the "Combined Set A" data set, which provided the best group of homogenous call-handling and time-documentation styles. This data set also contained complete response documentation from the time calls for aid were answered until ERUs were declared *en route* (see Sect. 5.2). This data set contained 22,564 fire responses and 115,206 EMS responses for analysis.

The same data set was used to examine the six questions with respect to the time of day factor, i.e., Daytime Versus Nighttime Response.

5.5.2 Daytime Versus Nighttime Response

To address this factor, the "Combined Set A" data set was separated into time-of-day categories based on the time the call was received. For purposes of this study, "Daytime" was defined as 0600 to 1800 h, accounting for 58 % of all responses, and "Nighttime" was defined as 0000–0600 hours, accounting for 15 % of all responses. Calls between the hours of 1800 and 0000 were analyzed but omitted from reporting for clarity. It was assumed that the daytime and nighttime groupings best represented the opposite ends of the spectrum when response personnel would be at their highest versus lowest daily levels of operational readiness. The distinction between fire and EMS calls was retained for this analysis. The data set provided 13,463 daytime and 2,681 nighttime fire responses and 66,202 daytime and 17,442 nighttime EMS responses.

5.5.3 Firefighter Crew Proficiency in Baseline Turnout Exercise

A separate data set was created for the data collected for the ideal time-to-task factor, namely, Firefighter Crew Proficiency in Baseline Turnout Exercise. This part of the study establishes a lower limit for Turnout Times based on simulating the component tasks common to turnout under ideal, controlled conditions. The data contained information from Baseline Turnout Exercise trials submitted by 13 departments. Overall, the data represented a total of 109 trials involving 327 career firefighters.

5.5.4 Effects of Station Layout on Turnout Response

The final data set contained layout information from the Station Information Sheets for the travel time within the ERF factor, i.e., Effects of Station Layout on Turnout Response. This part of the study establishes a basis for characterizing a "typical" firehouse in terms of foot-travel distances from common areas in the station to the ERU during an emergency turnout. Fourteen fire departments provided station layout data for a total of 225 fire stations.

Department Size as a Factor in Alarm Handling Time and Turnout Time

The response data available for this study was limited primarily to IAFC Class 5 fire departments; fire departments serving a population of 200,000 or more. This limits the direct extrapolation of these results to smaller departments.

Alarm handling is very much a "department-level" task. Based on five participating departments with similar call taking, dispatch, and documentation methods, we found a strong negative correlation between population served and median alarm handling time ($r = -0.78$) which suggests that larger departments typically process alarms more efficiently than smaller ones. One possible explanation for this observation is that very large departments by virtue of their size and economies of scale may tend more toward custom technology for improved automation of alarm handling. This is an area for future study.

Unlike alarm handling, turnout is more of a "personnel-level" task. Based on the same group of departments, we found a weak negative correlation between population served and median turnout time ($r = -0.22$) which suggests that larger departments turnout only slightly faster than smaller ones. A stronger indicator was the correlation between the average number of responses per station and turnout time ($r = 0.32$) which could suggest that rehearsal rather than budget is a factor in faster turnout times.

Chapter 6
Primary Findings

6.1 Fire Versus EMS Response

The most fundamental comparison made was between fire responses and EMS responses. We looked at

- *Alarm handling time*, a function of emergency dispatchers in the PSAP/Communication Center addressed by the NFPA 1221 standard.
- *Turnout time*, a function of fire and EMS crews within the ERF addressed by the NFPA 1710 standard.
- We bridged the two and looked at *mobilization time* overall to assess just how quickly the fire service, represented by our participant sample, actually "puts the rubber to the road" when the call for emergency aid is received (Fig. 6.1).

6.1.1 Alarm Handling Time

Benchmarks and Criteria
Both NFPA 1221 and NFPA 1710 use similar metrics to establish Call Processing Times and Turnout Times. The first part of the metric is the benchmark: a specified length of elapsed time. The second is the criteria: the percentage of responses within a sampling period that must occur at or below the benchmark time to achieve compliance with the standard.
For instance, part of the standard for Call Processing requires that 90 % of all responses (criterion) must be processed within 60 s (benchmark). Compliance with the standard can be measured in two ways:

R. Upson and K. A. Notarianni, *Quantitative Evaluation of Fire and EMS Mobilization Times*, SpringerBriefs in Fire, DOI: 10.1007/978-1-4614-4442-8_6,

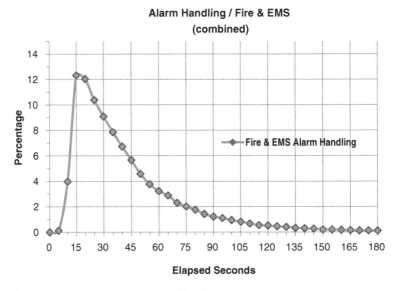

Fig. 6.1 Alarm handling/fire & EMS (combined)

1. Benchmark Compliance: What actual percentage of responses occurred at or below the benchmark time?
2. Criteria Compliance: How many seconds were actually required before the required percentage of responses occurred?

Tables throughout this document show the relevant NFPA benchmarks and criteria compared with the corresponding compliances recorded in the data.

Example: Widget Production Table

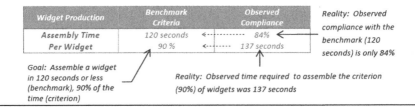

As stated previously, *alarm handling time* represents the elapsed time from the time a call for assistance is received at a PSAP, or "Alarm Time" (i.e., *Call intake;* when a 9-1-1 phone call is answered, when an automatic alarm is acknowledged, etc.), until appropriate ERUs are dispatched, or "Dispatch Time."

The findings of this study have been compared to the current NFPA 1221 standard, which sets two benchmark times with specific compliance criteria for Alarm handling:

Table 6.1 Alarm handling time/fire

Fire calls n = 22,564	NFPA 1221 benchmark criteria	Observed compliance	Median Mean Max
Alarm handling	60 s	79 %	29
	90 %	92 s	56
	90 s	90 %	3946
	99 %	315 s	

The confusing appearance of 90 % at both 90 and 92 s is an artifact of rounding. At 90 s, 89.72 % of all calls have been processed. It takes 92 s to reach 90.15 % of all calls processed

- 90 % of all emergency calls must be processed within 60 s or less.
- 99 % of all emergency calls must be processed within 90 s or less.

The standard makes no *alarm handling time* distinction between fire and EMS, but there are potentially significant differences in the nature of information, the amount of information, and the level of detail needed to accurately process Fire and EMS calls to warrant separate analyses. It could be argued that EMS responses, although they account for 84 % of all emergency responses in this study, are more likely to be triaged into less emergent response priority categories than fire responses, which have traditionally been categorized as emergencies. EMS has widely embraced call triage through standard *emergency medical dispatch* (EMD) protocols since the mid-1990 s, which could result in reduced processing times through formalization of the EMS call-taking process.

6.1.1.1 Fire

In the initial analysis, we examined 22,564 fire response records and noted that 79 % of all alarm handling for those observed responses was accomplished in 60 s or less with half of them accomplished in 29 s (median) or less (Table 6.1). This performance is well below the criterion set by NFPA 1221 for the 60 s benchmark. The mean average of 56 s is very close to the NFPA benchmark. Only 90 % of the responses were processed in 90 s or less, the second NFPA 1221 benchmark, as opposed to the 99 % in 90 s or less required by that standard (Fig. 6.2).

Regarding the time required to reach the criteria level required by the standard, it took 92 s to process 90 % of all fire response calls and 315 s—over 5 min—to process 99 % of all fire calls.

The maximum processing time for fire response was a matter of concern at well over an hour—clearly well beyond a typically acceptable processing time for an emergency call but nonetheless not inconceivable in a group of over 22,000 responses. The last 1 % of all responses analyzed ranged from over 315 to 3,946 s to process; 99 % of the responses required only 8 % of the observed range of values. It is likely that the processing times recorded in this extreme upper range represent grossly atypical responses, documentation errors, routine calls erroneously categorized as emergent, or some other form of data artifact. Several

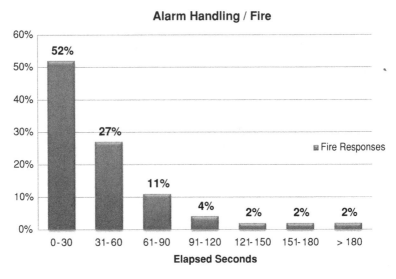

Fig. 6.2 Percent of fire alarms handled over time

Table 6.2 Alarm handling time/EMS

EMS calls n = 115,206	NFPA 1221 benchmark criteria	Observed compliance	Median Mean Max
Alarm handling	60 s	80 %	32
	90 %	84 s	44
	90 s	92 %	3565
	99 %	182 s	

methods for filtering out these extreme outliers were considered but rejected as moot, since it was determined that they had very little statistical impact on the overall analysis (see Appendix B) (Table 6.2).

6.1.1.2 EMS

Looking once more at the time that was actually needed to achieve the criteria required by the standard, it can be noted that it took 84 s to process 90 % of all EMS response calls and 182 s—about 3 min—to process 99 % of all EMS calls (Fig. 6.3).

The maximum processing time for an EMS response was slightly less than an hour. The time to process the last 1 % of the responses analyzed ranged between 182 and 3,565 s (59.4 min) to process: That 1 % of the responses accounted for 95 % of the observed range of values. It is likely that many of these *alarm handling times* represent the same type of artifacts presumed to be present in the

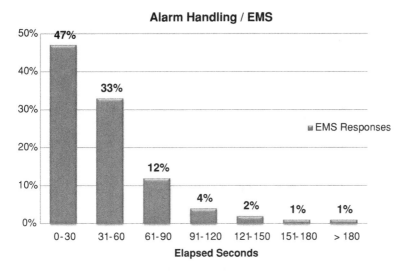

Fig. 6.3 Percent of EMS alarms handled over time

fire response data and they were again found to have very little statistical impact on the overall analysis. (see Appendix B)

6.1.1.3 Discussion Points

In the initial *alarm handling time* analysis, the fire and EMS responses were similar overall with respect to the current NFPA benchmarks. At the 60-second benchmark, about 80 % of all observed responses were processed, and nearly 90 % of all observed responses were processed at the 90-second benchmark. Alarm handling for fire responses does show a more skewed distribution of calls requiring more time to process than did the EMS responses. This distribution suggests that there may be a small qualitative difference between processing fire and EMS responses (Fig. 6.4).

The Cumulative Distribution Function (CDF)

CDFs in this document illustrate graphically the distribution of response data for alarm handling, turnout, and mobilization times. Starting at the lower left corner at 0 % and 0 s, a plot line representing the cumulative percentage of observed responses completed, shown on the vertical axis, is plotted against time elapsed to complete that percentage of responses, shown on the horizontal axis. The line rises quickly through the median average, 50 % of observed responses, and eventually passes through the various benchmark criteria that may be noted on the graph. The line "flattens out" quickly after the majority of responses have been completed and trails off in a long "tail" to the right as the last outliers are completed.

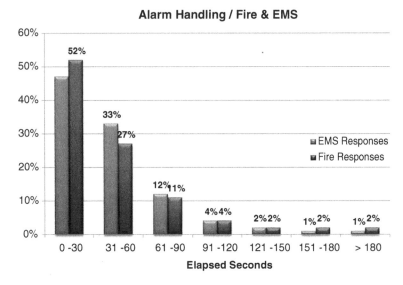

Fig. 6.4 Comparative percent of fire & EMS calls handled over time

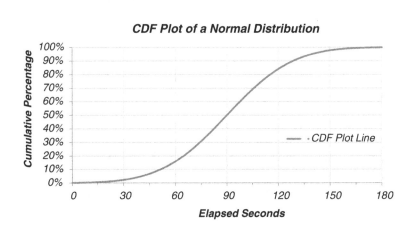

Using the above CDF plot of a normal distribution with a median value of 90 s, for example, the percentage of trials completed in 120 s or less can be determined. Starting at 120 s on the elapsed time x-axis, a vertical line would be drawn upward until it intersects the CDF plot. From there, a horizontal line is drawn to the left. The value at the point where that line intersects the cumulative percentage y-axis shows the percentage of trials completed in 120 s or less. In our example, for an elapsed time of 120 s, about 85 % of the trials are completed in 120 s or less.

Conversely, to answer the question of how long it would take to complete a certain percentage of calls, a similar process is followed in reverse. For

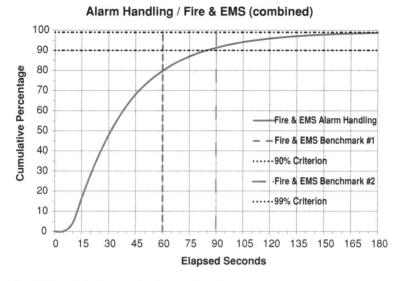

Fig. 6.5 CDF fire & EMS (combined) alarm handling time

example, in order to determine the time it would take to complete 90 % of all trials, one would start of the y-axis at the 90 % mark. A horizontal line would be drawn from this point to the right until the line intersects with the CDF plot. From this point of intersection, a vertical line is drawn down to the x-axis. The elapsed time read off the x-axis at this point is the answer. For our example, to process 90 % of the calls, requires a call processing time of about 130 s.

The previous edition of NFPA 1221 required a more stringent 95 % compliance time at the 60 s benchmark. With an 81 % observed compliance rate at 60 s, 95 % compliance would not be reached until ~106 s (Fig. 6.5). From the graph it can be seen that the cumulative distribution function crosses the 90 % mark very near to its upper inflection point. This is the point where the function begins to flatten faster than it rises. This point is arguably a more significant feature of the distribution to observe in terms of benchmark compliance than the 95 % mark.

Combining both fire and EMS alarm handling times into a single cumulative distribution function graph illustrates the sharp difference between the previous 95 % compliance criterion and present 90 % criterion. By the time the function achieves 90 % at around 83 s, it has reached its inflection point and is beginning to flatten more quickly than it rises. Beyond the 95 % mark the graph is flattening very quickly, which explains the lengthy time to compliance at the 99 % criterion.

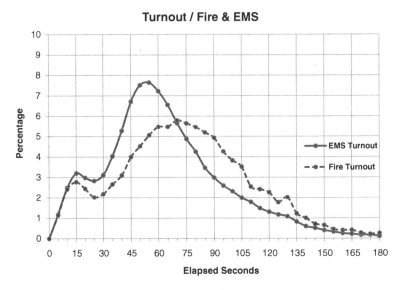

Fig. 6.6 Turnout/fire & EMS

6.1.2 Turnout Time

Turnout Time represents the elapsed time from the moment a call is dispatched, or "Dispatch Time" (i.e., when the call processer/dispatcher initiates an alert message to the assigned ERU.), until the assigned ERU(s) is physically en route, or "En route Time." The current NFPA 1710 standard sets separate benchmark times for fire and EMS responses with the same compliance criteria for Turnout (Fig. 6.6):

- 90 % of all emergency responses to fire calls must turnout within 80 s or less.
- 90 % of all emergency responses to EMS calls must turnout within 60 s or less.

Alarm handling is primarily a data-gathering operation, while turnout can be characterized primarily as a set of physical tasks. The typical tasks common to all turnouts from the ERF can reasonably be summarized as:

- Notification of the alarm
- Gathering critical response information
- Disengagement from tasks in process
- Travel within the ERF to the ERU
- Donning PPE
- Mounting the ERU and securing seatbelts
- Opening ERF bay doors
- Starting the ERU
- Signaling "en route"

For turnouts that originate outside of the ERF, when an ERU is already "on the air," the task list is considerably shorter:

- Notification of the alarm
- Gathering critical response information
- Signaling "en route"

The data contains turnout times for responses both from the ERU and "on the air" starts. The small peak in the response distributions shared by fire and EMS, about 15 s, is presumed to be representative of "on the air responses," while the much larger peaks represent normal responses from the ERU.

The NFPA 1710 standard makes a significant distinction between fire and EMS turnout time based on the slightly different tasks required as part of the turnout process. A response to a typical fire emergency requires donning structural fire-fighting PPE prior to mounting the ERU, whereas a response to a typical EMS call does not necessitate such extensive PPE.[1] Benchmarks for fire responses must accommodate additional turnout time to ensure that firefighters can safely don PPE before mounting the ERU. This permits seatbelts to be worn while en route in the interest of firefighter safety.

6.1.2.1 Fire

Analyzing a set of 22,564 fire response records (Fig. 6.10), we noted that only 60 % of all recorded turnouts were accomplished in 80 s or less with half of them accomplished in 71 s (median) or less. This result is a well below the performance criterion set by NFPA 1710 for fire responses. The mean average of 75 s is very close to the NFPA benchmark (Fig. 6.7).

Looking at the time actually needed to reach the criteria required by the standard, it can be noted that it took 123 s to reach the 90 % criterion for reported fire call responses—over one and one-half times the time allotted by the standard benchmark (Table 6.3).

The maximum reported turnout time for fire response ranged up to 45 min. The extreme outliers within this range presumably contain some mixture of atypical actual turnout times and documentation artifacts. As with alarm handling, it was determined that the extreme outliers, representing 1 % of the data and 95 % of the range of reported values, had very little statistical impact on the overall analysis.

---

[1] "…This is believed to be due to the fact that dressing in structural firefighting protective clothing prior to boarding the fire apparatus takes more time…. Because fire fighters do not need to dress in structural firefighting protective clothing for EMS responses, the extra 20 s of turnout time was not felt to be necessary for these responses" (NFPA 1710 ROC 2009, 1710-5).

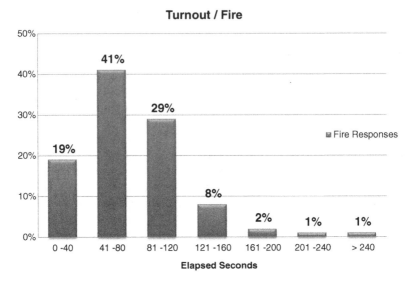

Fig. 6.7 Percent of fire turnouts completed over time

Table 6.3 Turnout time/fire

Fire calls n = 22,564	NFPA 1710 benchmark criteria	Observed compliance	Median Mean Max
Turnout	80 s 90 %	60 % 123 s	71 75 2629

Table 6.4 Turnout time/EMS

EMS calls n = 115,206	NFPA 1710 benchmark criteria	Observed compliance	Median Mean Max
Turnout	60 s 90 %	54 % 109 s	58 63 3112

6.1.2.2 EMS

Analysis of 115,206 EMS response records, summarized in Table 6.4, shows only 54 % of all recorded turnouts were accomplished in the more stringent 60 s or less required for EMS with half of them accomplished in 58 s (median) or less. This is below the performance criterion set by NFPA 1710 for EMS call responses. Once again, the weighted average of 63 s (mean) is very close to the NFPA benchmark (Fig. 6.8).

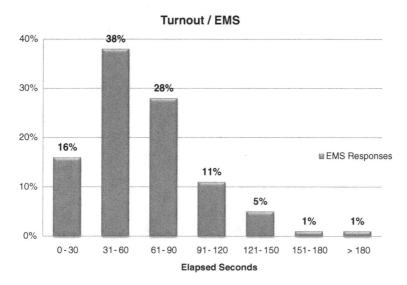

Fig. 6.8 Percent of EMS turnouts completed

Looking at the time required to reach the compliance criteria set by the standard, it is noted that it took 109 s to account for 90 % of turnouts to EMS call responses—over one and two-thirds times the time allotted by the standard benchmark.

The maximum reported turnout time for EMS response was nearly 52 min. As was observed with fire call response turnout, it was determined that the outliers, representing 1 % of the data and 96 % of the range of reported values, had very little statistical impact on the overall analysis.

6.1.2.3 Discussion Points

In the overall *turnout time* analysis, we found the recorded fire and EMS responses were generally consistent with the benchmarks set for each in the NFPA 1710 standard. The recorded fire response turnouts required, on average, 12 s longer than EMS response turnouts with 54–60 % of all recorded responses recording turnout times at or below the appropriate standard benchmark. Turnout times for both fire and EMS responses required 43–49 s beyond the standard benchmarks to reach the 90 % criterion. This result suggests that the standard may be underestimating the time it takes to complete the baseline turnout tasks common to both fire and EMS responses in establishing the benchmark. Section 6.3 Firefighter Crew Proficiency in Baseline Turnout Exercise, examines the ideal baseline time-to-task measurement for fire call responses.

The previous edition of NFPA 1710 did not set a separate benchmark for fire and EMS responses. In that version, all responses shared a common 60-s

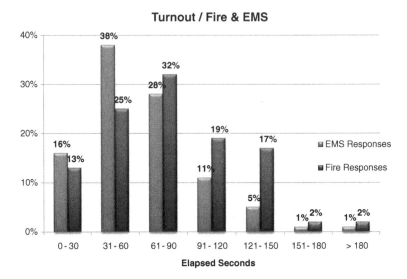

Fig. 6.9 Comparative percent of fire & EMS turnout completed over time

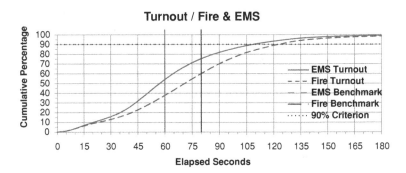

Fig. 6.10 CDF fire & EMS turnout time

benchmark. Less than 40 % of the recorded fire call responses showed turnout times of 60 s or less (Fig. 6.9). From the graph it can be seen that the cumulative distribution function crosses the 60 s mark early in its rise. This is well below its upper inflection point. An arguably more significant point to observe in terms of fire response turnout benchmark compliance may be the apparent inflection point around 120 s, which corresponds closely with the 90 % mark (Fig. 6.10).

Likewise, in terms of EMS response turnout, benchmark compliance suggested in Fig. 6.10 may be at the apparent inflection point at 110 s, corresponding closely with the 90 % mark.

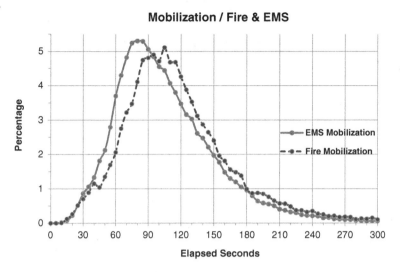

Fig. 6.11 Mobilization/fire & EMS

Table 6.5 Mobilization time/fire

Fire calls n = 22,564	Implicit benchmark criteria	Observed compliance	Median Mean Max
Mobilization	140 s	75 %	108
	81 %	154 s	130
			5966

6.1.3 Mobilization Time

Mobilization time brings together the complete process of receiving the call for aid at "Alarm Time;" determining and assigning appropriate ERUs at "Dispatch Time;"and getting those ERUs on the road to the scene of the emergency at "En route Time." Combining performance criteria in NFPA 1221 and NFPA 1710 standards yields implicit fire and EMS benchmark times with a common performance criterion (Fig. 6.11):

- 81 % of all emergency responses to fire calls must turnout within 140 s or less.
- 81 % of all emergency responses to EMS calls must turnout within 120 s or less.

As we have seen, the compliance rates for *alarm handling time* and *turnout time* were not observed in practice to be as high as their respective standards required. However, the combined rates have the advantage of a lower standard that is implied by the statistical combination of the performance criteria in the NFPA 1221 and NFPA 1710 standards.

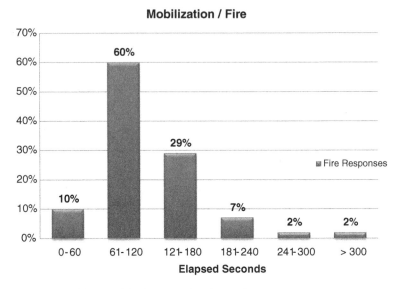

Fig. 6.12 Percent of fire mobilizations completed over time

Table 6.6 Mobilization time/EMS

EMS calls n = 115,206	Implicit benchmark criteria	Observed compliance	Median Mean Max
Mobilization	120 s 81 %	70 % 141 s	96 107 3615

6.1.3.1 Fire

More than 22,500 records for recorded *mobilization times* for fire responses (Table 6.5) show a benchmark compliance rate of 75 %. Half of the recorded mobilizations turnouts (median) were completed in 108 s or less. This is only 6 % below the implied mobilization for fire call responses. The weighted average of 130 s (mean) is only 10 s below the implied benchmark.

Looking at how much time was actually needed to reach the performance criterion required by the standard, it is noted that it took 154 s to reach the 81 % criterion for reported fire call responses—only 14 s longer than the time allotted by the implied benchmark (Fig. 6.12).

The maximum reported mobilization time for fire response was over one and one-half hours. This is likely the result of extreme outliers in both *alarm handling times* and *turnout times*. Although the effects of these extreme outliers continue to show themselves in the uppermost cumulative percentages, they remain statistically insignificant overall.

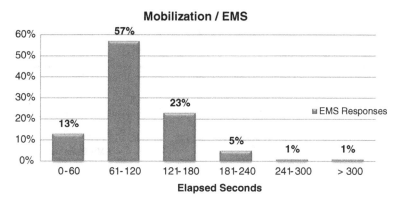

Fig. 6.13 Percent of EMS mobilizations completed over time

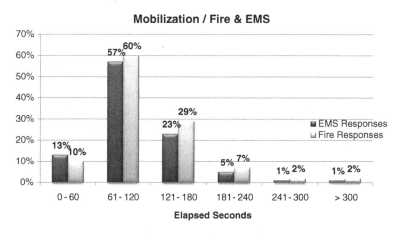

Fig. 6.14 Comparative percent of fire & EMS mobilization completed over time

6.1.3.2 EMS

More than 115,000 records for EMS response *mobilization times* (Table 6.6) show a benchmark compliance rate of 70 %. Half of the recorded mobilizations turnouts (median) were completed in 107 s or less (only 1 s less than fire responses). This result is 11 % below our implied mobilization criterion for fire call responses, which places EMS responses slightly below the fire responses in overall mobilization criterion compliance. Note that the weighted average of 107 s (mean) is only 13 s below the implied benchmark.

Looking at how much time was actually needed to achieve the performance criteria required by the standard, it is noted that it took 141 s to reach the 81 %

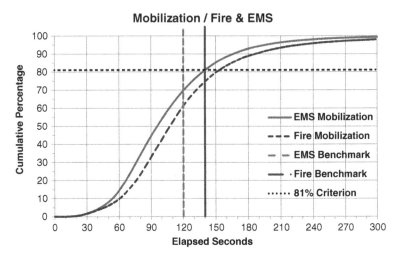

Fig. 6.15 CDF fire & EMS mobilization time

criterion for reported fire call responses—21 s longer than the time allotted by the implied benchmark (Fig. 6.13).

The maximum reported mobilization time for EMS response ranged to just over 1 h. As seen previously with fire responses, this is the result of extreme outliers in both alarm handling and turnout times. Although the effects of these outliers continue to show themselves in the uppermost cumulative percentages, they remain statistically insignificant overall.

6.1.3.3 Discussion Points

With respect to a hypothetical 90 % criterion equivalent to that imposed on *alarm handling time* and *turnout time*, an additional 30 + seconds would be required beyond the current benchmarks for both Fire (187 s required) and EMS (167 s required) (Fig. 6.14). Cumulative distribution functions for both fire and EMS mobilization show similar curves for fire and EMS (Fig. 6.15), with fire responses showing a longer "tail." This may indicate a more common pairing of both longer than average *alarm handling times* and longer than average *turnout times*.

6.2 Daytime Versus Nighttime Response

Data for responses were divided into Daytime (responses for which the alarm was received between 0600 and 1800 h) and Nighttime (responses for which the alarm was received between 0000 and 0600 h) and analyzed for compliance with the NFPA standards.

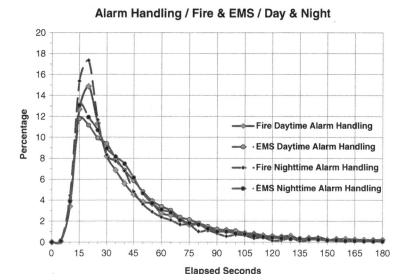

Fig. 6.16 Alarm handling/fire & EMS/day & night over time

Table 6.7 Alarm handling time/fire/daytime

Fire calls n = 13,463	NFPA 1221 benchmark criteria	Observed compliance	Median Mean Max
Alarm handling	60 s	78 %	30
	90 %	98 s	64
	90 s	88 %	3946
	99 %	475 s	

6.2.1 Alarm Handling Time

NFPA 1221 makes no distinction between daytime and nighttime *alarm handling time*, and the benchmarks used are the same as those compared previously in Table 6.1. Alarm Handling Time/Fire and Table 6.2. Alarm Handling Time/EMS. Those two benchmarks and compliance for alarm handling criteria are (Fig. 6.16):

- 90 % of all emergency calls must be processed within 60 s or less.
- 99 % of all emergency calls must be processed within 90 s or less.

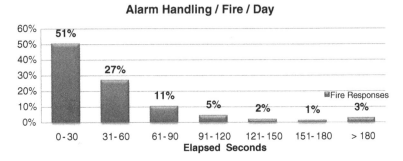

Fig. 6.17 Percent of daytime fire calls handled over time

Table 6.8 Alarm handling time/fire nighttime

Fire calls n = 2,681	NFPA 1221 Benchmark Criteria	Observed compliance	Median Mean Max
Alarm handling	60 s	85 %	26
	90 %	73 s	39
	90 s	94 %	1029
	99 %	183 s	

6.2.1.1 Fire/Daytime

Daytime responses accounted for 58 % of all responses analyzed. The recorded compliance rate for daytime fire responses is very similar to the overall rate from Table 6.1. Alarm Handling Time/Fire. There is a slight drop in compliance, with only 78 % of 13,463 recorded fire calls processed in 60 s or less compared to the earlier reported 80 % overall (Table 6.7). The number of calls processed in 90 s or less also drops 2–88 % with a substantial increase in the number of seconds needed to process the required 99 % of responses. This can likely be attributed to the range of observed processing times, which is greatest in this segment with a maximum alarm handling time of almost 66 min.

An explanation behind for maximum *alarm handling time* of well over an hour for fire response calls remains a mystery. However, for the data analyzed, it appears that such long processing times tend to occur during the daytime rather than nighttime (Fig. 6.17).

6.2.1.2 Fire/Nighttime

There is notable improvement in compliance for the 2,681 recorded nighttime calls. Alarm handling during this period shows a lower median time, at just 26 s, and a substantially lower mean time, 39 s. This may be driven by a much more narrow range

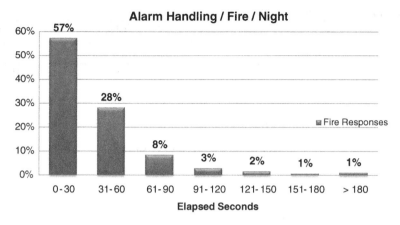

Fig. 6.18 Percent of nighttime fire calls handled over time

Table 6.9 Alarm handling time/EMS/daytime

EMS calls n = 66,202	NFPA 1221 benchmark criteria	Observed compliance	Median Mean Max
Alarm handling	60 s	79 %	33
	90 %	85 s	44
	90 s	91 %	3565
	99 %	182 s	

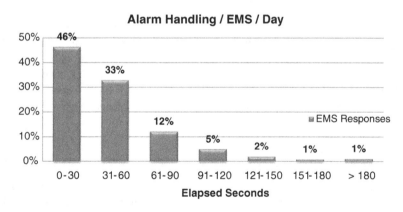

Fig. 6.19 Percent of daytime EMS calls handled over time

of observed processing times (Table 6.8). For this time segment, observed *alarm handling time* comes very close to compliance with the NFPA 1221 alarm handling criteria, with 85 % of all calls processed within 60 s and 94 % in 90 s (Fig. 6.18).

Table 6.10 Alarm handling time/EMS/nighttime

EMS calls n = 17,442	NFPA 1221 benchmark criteria	Observed compliance	Median Mean Max
Alarm handling	60 s	81 %	31
	90 %	80 s	42
	90 s	93 %	1532
	99 %	181 s	

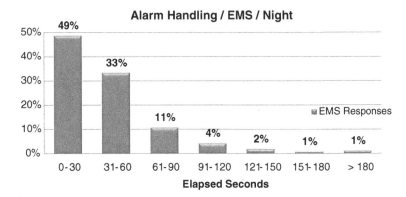

Fig. 6.20 Percent of nighttime EMS calls handled over time

6.2.1.3 EMS/Daytime

As noted for alarm handling for Fire/Daytime responses, the EMS compliance rate is very similar to the overall rate from Table 6.2. Alarm Handling Time/EMS, down only 1–79 % of all calls processed within 60 s and an additional 1–91 % for calls processed at or under 90 s (Table 6.9). The wide range of observed processing times noted in Fire/Daytime appears once again in EMS. This result raises a question regarding what factors contribute to this difference (Fig. 6.19).

6.2.1.4 EMS/Nighttime

In contrast to daytime *alarm handling time*, recorded nighttime criteria compliances increased by one percent each from their overall values to 81 % of calls processed at or below 60 s and 93 % processed within 90 s. This result was accompanied by a much more narrow range of observed processing times (Table 6.10).

There is no explanation for the maximum processing time for EMS recorded responses, slightly less than 1 h, in the data. Further research may offer some explanation of why *alarm handling time* tends toward more extreme outliers during the daytime for both fire and EMS calls (Fig. 6.20).

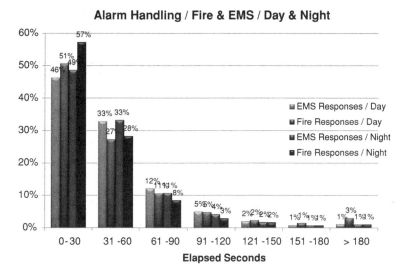

Fig. 6.21 Comparative percent of day & night fire & EMS calls handled over time

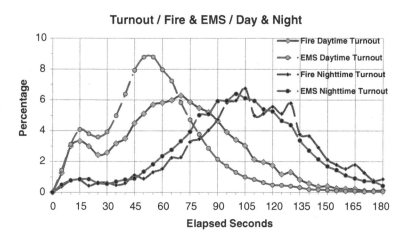

Fig. 6.22 Turnout/fire & EMS/day & night over time

6.2.1.5 Discussion Points

In the analysis of daytime versus nighttime alarm handling, there were unexpected differences in *alarm handling times* that are not addressed in the NFPA 1221 standard. *Alarm handling times* for both fire and EMS responses were typically completed in less time during the nighttime period than daytime (Fig. 6.21).

Table 6.11 Turnout time/fire/daytime

Fire calls n = 13,463	NFPA 1710 benchmark criteria	Observed compliance	Median Mean Max
Turnout	80 s	67 %	66
	90 %	112 s	68
			2629

- Benchmark compliance for alarm handling for fire calls improved from 78 % during the day to 85 % at night.
- Benchmark compliance for alarm handling for EMS calls improved from 81 % during the day to 83 % at night.
- The largest elapsed alarm handling times occurred during the day.
- Are there differences in the range of processing complexity for day versus night calls for aid?

  – Are the calls different?
  – Are the callers different?

6.2.2 Turnout Time

The current NFPA 1710 standard does not address the difference in day or night turnout using separate benchmark times.[2] The turnout standard for fire and EMS responses, regardless of time of day, is (Fig. 6.22):

- 90 % of all emergency responses to fire calls must turnout within 80 s or less.
- 90 % of all emergency responses to EMS calls must turnout within 60 s or less.

The tasks common to all turnouts from the ERF were summarized in the previous section and still apply here with changes and additions accounting for nighttime activity:

- Notification of the alarm
- Gathering critical response information
- Disengagement from tasks in process

  – Disengagement from task in progress may now include waking and orienting
  – Dressing to the level of station wear required by local practice

- Travel within the ERF to the ERU

  – Detour to restroom

---

[2] "The committee does not see the need to establish separate turnout times by time of day. Fire departments that experience significant differences depending on the time of day should evaluate what is going on during those periods and determine if there are ways to improve those response times." (NFPA 1710 ROC 2009, 1710–5).

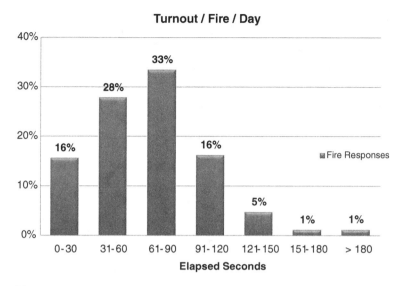

Fig. 6.23 Percent of daytime fire turnouts completed over time

Table 6.12 Turnout time/fire/nighttime

Fire calls n = 2,681	NFPA 1710 benchmark criteria	Observed compliance	Median Mean Max
Turnout	80 s	21 %	108
	90 %	158 s	110
			1058

- Donning PPE
- Mounting the ERU and securing seatbelts
- Opening ERF bay doors
- Starting the ERU
- Signaling "en route"

6.2.2.1 Fire/Daytime

Analysis of a set of 13,463 daytime fire response records (Table 6.11) revealed that two-thirds (67 %) of all recorded daytime turnouts were accomplished in 80 s or less, with half of them accomplished in 66 s (median) or less. This is below the criterion set by NFPA 1710 for fire responses, and the mean average of 68 s is below the NFPA benchmark.

Looking at how much time was actually needed to achieve the performance criteria required by the standard, it is noted that it took 112 s to reach the 90 % criterion for reported fire responses. This is slightly more than one and one-third times the time allotted by the standard benchmark.

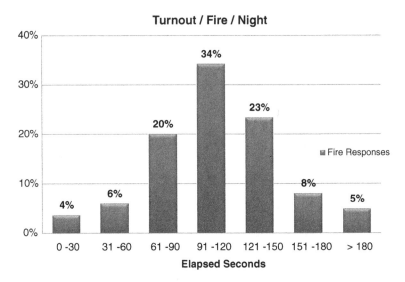

Fig. 6.24 Percent of nighttime fire turnouts completed over time

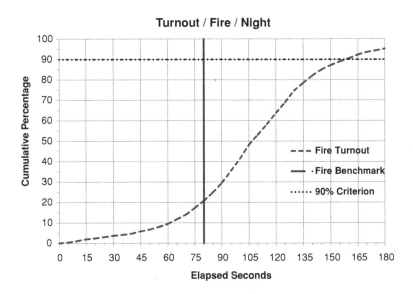

Fig. 6.25 CDF fire turnout time/nighttime

The 45 min maximum reported turnout time and other extreme outliers noted in the overall analysis (Table 6.3. Turnout Time/Fire) appear during the daytime hours continuing the trend of outliers occurring during the day (Fig. 6.23).

Table 6.13 Turnout time/EMS/daytime

EMS calls n = 66,202	NFPA 1710 benchmark criteria	Observed compliance	Median Mean Max
Turnout	60 s	65 %	52
	90 %	87 s	54
			3112

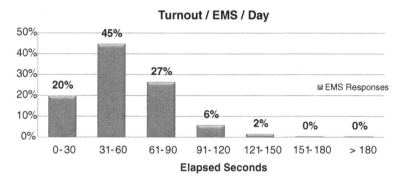

Fig. 6.26 Percent of daytime ems turnouts completed over time

6.2.2.2 Fire/Nighttime

There was a significant drop in benchmark compliance during the 2,681 recorded nighttime responses (Table 6.12). Only 21 %, about one response in five, met the 80-second benchmark. The median turnout response during this period took 108 s, and 158 s are required to achieve the standard's 90 % compliance criterion (Fig. 6.24).

Examining the cumulative distribution function shown in Fig. 6.25, it can be noted that the main body of responses, represented by the steepest part of the curve, has not begun by the time the benchmark has passed. The majority of turnouts do not occur until between 80 s and 150 s.

6.2.2.3 EMS/Daytime

Daytime EMS responses totaled 66,202 (Table 6.13). For this group, just under two-thirds (65 %) of all recorded daytime turnouts were accomplished in 60 s or less, with half of them accomplished in 52 s (median) or less. This is below the criterion set by NFPA 1710 for EMS responses, and the weighted average of 54 s (mean) is close to the NFPA benchmark.

Looking at how much time was actually needed to achieve the criteria required by the standard, it can be shown that it took 87 s to reach the 90 % criterion for

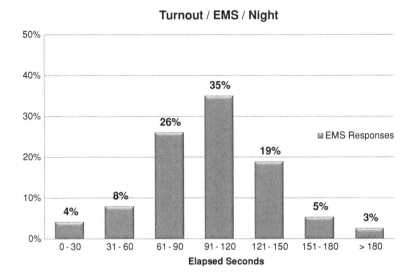

Fig. 6.27 Percent of nighttime EMS turnouts completed over time

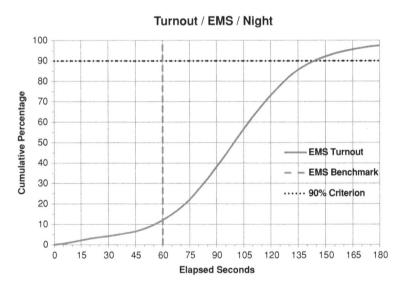

Fig. 6.28 CDF EMS turnout time/nighttime

reported fire call responses. This is almost one and one-half times the time allotted by the standard benchmark.

The 52 min maximum time reported and other extreme outliers noted in the overall analysis (Table 6.4. Turnout Time/EMS) appears during the daytime hours, continuing the trend of outliers during the day (Fig. 6.26).

Table 6.14 Turnout time/EMS/nighttime

EMS calls n = 17,442	NFPA 1710 benchmark criteria	Observed compliance	Median Mean Max
Turnout	60 s	12 %	100
	90 %	144 s	101
			2142

Table 6.15 Turnout time/fire vs EMS

Turnout	Daytime		Nighttime		Δ	
Seconds	Median	90 %	Median	90 %	Median	90 %
Fire	66	112	108	158	+42	+46
EMS	52	87	100	144	+48	+57
Δ	-14	-25	-8	-14		

6.2.2.4 EMS/Nighttime

Nighttime turnout compliance drops considerably in the 17,442 recorded EMS responses than it did in fire responses (Table 6.14). With only 12 % of recorded responses completing turnout in 60 s or less—about one response in eight—the average nighttime turnout response takes 101 s, and 144 s are required to achieve the standard's 90 % compliance criterion (Fig. 6.27).

Examining the cumulative distribution function shown in Fig. 6.28, it can be noted that the responses have only begun when the benchmark has passed. The majority of turnouts occur between 60 s and 140 s.

6.2.2.5 Discussion Points

The analysis of daytime versus nighttime response turnout noted differences not specifically addressed in the NFPA 1710 standard. Nighttime turnout times for both fire and EMS responses were significantly below the standard benchmarks:

- 80 s benchmark compliance for Turnout Time for fire responses decreased from 67 % during the day to 21 % at night.
- 60 s benchmark compliance for Turnout Time for EMS responses decreased from 65 % during the day to 12 % at night.
- Conversely, the largest elapsed Turnout Times occurred during the day.

Observed turnout times for Fire and EMS during the daytime period, compared in Table 6.15, show an average difference of 14 s (medians) to 25 s (time to 90 % criterion). This is commensurate with the 20-second allowance in different benchmarks for Fire and EMS set by NFPA 1710. Average nighttime period turnouts, which increase from daytime turnouts by similar increments for both fire

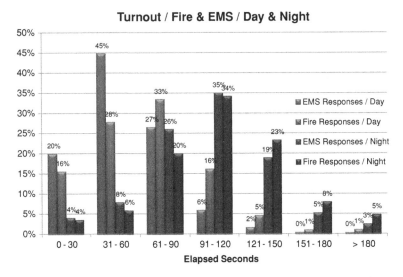

Fig. 6.29 Comparative percent of day & night fire & EMS turnouts completed over time

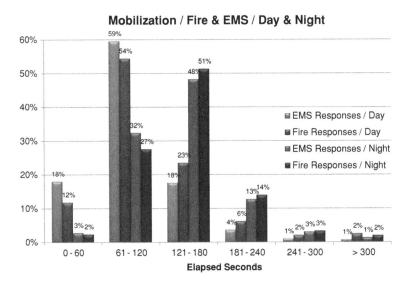

Fig. 6.30 Percent of day & night fire & EMS mobilizations completed over time

and EMS, become similar, with an average difference of only 8 s (medians) to 14 s (time to 90 % criterion). The fact that the two types of responses become more aligned as they increase in elapsed time suggests that the main variation between fire and EMS turnout—different PPE requirements—becomes less important as new common tasks responsible for 42–48 s (fire and EMS median Δ) of turnout time are added during the nighttime turnout response (Fig. 6.29).

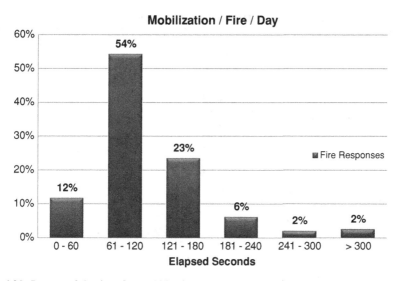

Fig. 6.31 Percent of daytime fire mobilizations completed over time

Table 6.16 Mobilization time/fire/daytime

Fire calls n = 13,463 (328)	Implicit benchmark criteria	Observed compliance	Median Mean Max
Mobilization	140 s	78 %	103
	81 %	148 s	134
	90 %	184 s	5966

Table 6.17 Mobilization time/fire/nighttime

Fire calls n = 2,681 (125)	Implicit benchmark criteria	Observed compliance	Median Mean Max
Mobilization	140 s	52 %	138
	81 %	180 s	149
	90 %	208 s	1671

6.2.3 Mobilization Time

Because the two response segments, *alarm handling time* and *turnout time*, are essentially independent of each, it is unlikely that the extremes of each would coincide in the same response incident, though it could occur. Therefore when the data for these two response segments were combined, there was near compliance with the combined standard benchmarks during daytime responses (Fig. 6.30).

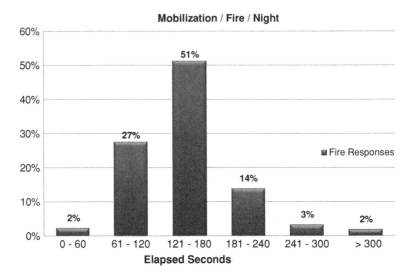

Fig. 6.32 Percent of nighttime fire mobilizations completed over time

Table 6.18 Mobilization time/EMS/daytime

EMS calls n = 66,202	Implicit benchmark criteria	Observed compliance	Median Mean Max
Mobilization	120 s	77 %	88
	81 %	127 s	99
	90 %	153	3615

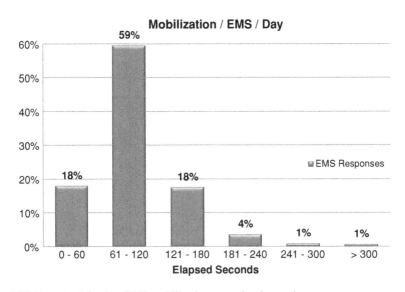

Fig. 6.33 Percent of daytime EMS mobilizations completed over time

Table 6.19 Mobilization time/EMS/nighttime

EMS calls n = 17,442 (822)	Implicit benchmark criteria	Observed compliance	Median Mean Max
Mobilization	120 s	35 %	135
	81 %	176 s	143
	90 %	203 s	2219

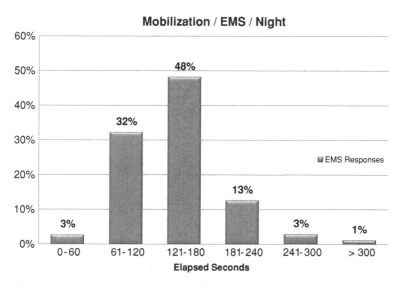

Fig. 6.34 Percent of nighttime EMS mobilizations completed over time

6.2.3.1 Fire/Daytime

As noted, daytime compliance is slightly better than overall compliance as shown in Table 6.5. (Fig. 6.31; Table 6.16).

6.2.3.2 Fire/Nighttime

Mobilization compliance (Table 6.17) drops considerably during the nighttime period to only 55 % compliance with the combined performance criterion created from the criteria in NFPA 1221 and NFPA 1710, and required 32 s beyond the benchmark to achieve compliance (Fig. 6.32).

Table 6.20 Turnout exercise summary

Turnout exercise n = 106	NFPA 1710 benchmark criteria	Observed compliance	Median Mean Max
En route	80 s	81 %	68
"Wheels rolling"	90 %	85 s	67
			112
En route	80 s	70 %	72
"Crosses sill"	90 %	92 s	74
			114

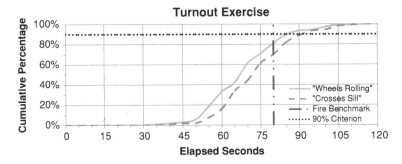

Fig. 6.35 CDF baseline turnout exercise

6.2.3.3 EMS/Daytime

Daytime compliance (Table 6.18) improved by 7 % compared to overall compliance as shown in Table 6.6 Mobilization Time/EMS. As occurred with Fire/Daytime responses, compliance with the implied criterion is very nearly achieved (Fig. 6.33).

6.2.3.4 EMS/Nighttime

Mobilization compliance (Table 6.19) drops considerably during the nighttime period to only 37 %, with the implied criterion requiring 48 s beyond the benchmark to achieve compliance (Fig. 6.34).

6.3 Firefighter Crew Proficiency in Baseline Turnout Exercise

Results from a total of 106 turnout exercises were submitted, representing 13 participating fire departments. Since there is no definition provided in NFPA 1710 and no clear peer consensus of criteria on "en route" status, results were collected

Table 6.21 Alarm response/horizontal

Alarm response walk (Horizontal travel)	Mean walking speed (fps)	Seconds per 50'
N = 335	4.98	10.03

and are reported using both the moment the apparatus' wheels were visibly rolling[3] ("wheels rolling") and the moment when the front bumper of the apparatus crossed the garage door sill[4] ("crosses sill") (Table 6.20).

The results indicate that, even under ideal conditions, the process of turnout requires substantially longer than the NFPA 1710 standard currently allows. Using the more liberal "wheels rolling" criterion, benchmark compliance is achieved only 81 % of the time rather than the 90 % target established by the standard. In order to reach the 90 % target, 85 s were required (Fig. 6.35). Using the more conservative "crosses sill" criterion, benchmark compliance is achieved only 70 % of the time, with 92 s required to achieve the targets established by the standard.

These results are slightly slower but consistent with the preliminary results of the original study utilizing the Baseline Turnout Exercise.[5]

6.4 Effects of Station Layout on Turnout Response

The physical attribute of an ERF has a direct influence on firefighter *turnout time*. Responding crews must traverse between work and other activity areas and the ERU itself. Both horizontal and vertical foot travel distances add time to any emergency turnout.

6.4.1 Horizontal Travel

A commonly cited factor for calculating travel times by average adults without a locomotor disability is a mean walking speed of 4.10 fps (*1.25 m/s*). This factor has been measured for horizontal travel while evacuating a building (Boyce, Shields and Silcock 1999, 54). This measure is cited in both *The SFPE Handbook of Fire Protection Engineering* (Bryan Behavioral Response to Fire and Smoke 2002, 3–

---

[3] This criterion has been suggested informally in conversation with members of the NFPA 1710 committee.

[4] The "crosses sill" criterion is suggested in *NFPA Structural Firefighting Strategy and Tactics*: "The third segment is the turnout time. This is the time from the receipt of the alarm until the apparatus crosses the front door sill of the station."(Klaene and Sanders 2008, 125).

[5] With n = 38, mean "en route" times of 68 and 70 s respectively were recorded for "wheels rolling" and "crosses sill" criteria. A third criteria, "rear bumper crosses sill" (mean "en route" time of 74 s) was dropped from the standardized version of the exercise for this study to make it more portable (Upson 2009).

Table 6.22 Alarm response/vertical

Alarm response (vertical travel)	Down (seconds)	Down (fps)	Up (seconds)	Up (fps)
n	11	11	10	10
Straight Run Stair (8.5′)	3.46	2.46	2.94	2.89
Return Run Stair (10′)	6.53	1.53	6.16	1.62
Fire Pole (10′)	5.52			
Estimated Typical				2.13

Table 6.23 Alarm response/conversion

Horizontal equivalency	Horizontal travel (fps)	Vertical travel (fps)	Conversion (Horizontal/ Vertical)	Horizontal equivalent nominal 10′ stair (feet)
	4.98	2.30	2.17	21.7

Table 6.24 Station layout summary

Station layout n = 197	Day room travel n = 195	Training room travel n = 179	Dining/kitchen travel n = 195	Fitness room travel n = 190	Sleeping room travel n = 197
min	11	6	8	5	13
median	72	72	69	68	75
mean	70	71	71	74	84
std dev	34	37	30	43	36
max	192	237	155	226	212

360) and in the *Fire Protection Handbook* (Bryan, Human Behavior and Fire 2008, 4–40). For this study a factor closer to 5.8 fps, the highest speed recorded by Boyce, Shields and Silcock, was assumed to be more appropriate for firefighters moving to their ERU during turnout.

To empirically assess this factor, firefighters from 13 participating fire departments were timed over a measured indoor course walking as if they were responding to an emergency call. Firefighters participating in the timed walking exercise were instructed "not to run." This exercise established a reasonable estimation of how quickly firefighters might be expected to travel safely to reach their apparatus when actually responding to an alarm.

The mean walking speed recorded was 5 fps, or 10 s for every 50 feet of horizontal foot travel, as shown in Table 6.21. This exercise, conducted by multiple raters in 13 fire departments, is slightly slower but not inconsistent with the results of an earlier study that recorded an average speed of 5.7 fps using a highly

---

[6] With n = 8, that study cites a mean descending speed of 2.3 fps with a range of 1.5–3.6 fps and an interquartile range of 1.7–2.9 fps.

[7] Candidate Physical Ability Test Program (The IAFF/IAFC Wellness Fitness Task Force n.d.).

motivated and competitive subject pool associated with the DHS-funded Fire-fighter Safety and Deployment Study (Upson 2009).

6.4.2 Vertical Travel

A smaller, more closely controlled set of timed exercises was used to estimate firefighter travel times for travel up and down stairs and down fire poles. A conservative value of 1.71 fps for vertical travel was chosen to represent a reasonable estimation of all typical vertical indoor travel. This value is actually slower than the value of 2.3 fps cited by Boyce, Shields and Silcock (Boyce, Shields and Silcock 1999) but is within the lower average range.[6] Firefighters participating in timed stair exercises were instructed not to run and to "touch every step" as a safety measure consistent with instructions used in CPAT program,[7] which conceivably resulted in more conservative average speeds (Table 6.22).

In order to more easily quantify travel time using combined horizontal and vertical components, a horizontal equivalency was calculated for a nominal 10-foot flight of stairs. Based on the mean horizontal and vertical speeds noted above, a conversion factor of 2.17 was derived. This is equivalent to 22 feet of horizontal travel for each nominal 10 feet of vertical travel (Table 6.23).

6.4.3 Station Layout

The Baseline Turnout Exercise is based on an event in which all the crew members are within 50 feet of their assigned apparatus at the time of alarm. To estimate times in which foot travel exceeds 50 feet, a walking speed of 5 fps,[8] or 10 s for every additional 50 feet of travel, can reasonably be used to project the minimum turnout time required.

Measurements of horizontal travel distances were made from the door of the primary apparatus to various key locations in 197 fire stations. Where vertical components were part of the path of travel, the horizontal equivalency calculated above was added to the horizontal distance. Based on the average travel distances recorded, it is reasonable to assume that firefighters responding to an alarm may typically have to travel in excess of 100 feet inside the station to reach their assigned ERU. This measure equates to another 10 s of turnout time beyond the baseline established by the Ideal Turnout Exercise (Table 6.24).

---

[6] With n = 8, that study cites a mean descending speed of 2.3 fps with a range of 1.5–3.6 fps and an interquartile range of 1.7–2.9 fps.

[7] Candidate Physical Ability Test Program (The IAFF/IAFC Wellness Fitness Task Force n.d.).

[8] With n = 131, mean walking speed was calculated at 5.7fps (Upson 2009).

Chapter 7
Conclusions

I. The actual recorded *alarm handling times*, provided to this study from a group of large fire departments, were compiled, statistically analyzed, and compared to the target alarm handling times given in NFPA 1221. Results demonstrated that:

    a. For both fire and EMS calls, the mean average alarm handling times observed were less than 60 s.

    b. For approximately 80 % of the fire and EMS calls, alarm handling was completed in the required 60 s or less.

    c. Eighty percent of calls processed in 60 s or less falls below the 90 % targeted in the standard.

    d. The time required for alarm handling of 90 % of the calls was 92 s for fire (slightly over one and one-half times the standard) and 84 s for EMS (slightly less than one and one-half times the standard).

    e. A second benchmark is set in the standard, which targets 90 s to process 99 % of the calls. At an elapsed time of 90 s, approximately 90 % of the calls were processed rather than the 99 % required. Given the observed distribution of alarm handling times, where a very long tail is observed, the 99 % criterion may not be particularly useful for benchmarking. A long tail is observed in the distribution, which represents long alarm handling times for a certain fraction of the fire and EMS calls.

R. Upson and K. A. Notarianni, *Quantitative Evaluation of Fire and EMS Mobilization Times*, SpringerBriefs in Fire, DOI: 10.1007/978-1-4614-4442-8_7,
© Fire Protection Research Foundation 2010

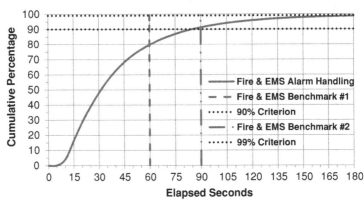

II. The actual recorded *turnout times*, provided to this study from a group of large fire departments, were compiled, statistically analyzed, and compared with the target alarm handling times given in NFPA 1710.

  a. For both fire and EMS calls, the mean average turnout times observed fell well within their respective current benchmarks of 80 s for fire and 60 s for EMS.

   i. For approximately 60 % of the fire calls, turnout was completed in the required 80 s or less.
   ii. For approximately 54 % of the EMS calls, turnout was completed in the required 60 s or less.

  b. The time actually required and recorded for turnout of 90 % of the calls was 123 s for fire (slightly over one and one-third times the standard) and 109 s for EMS (slightly more than one and two-thirds times the standard).

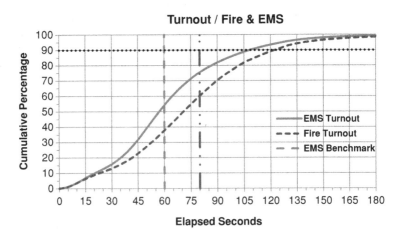

III. The actual recorded *turnout times*, provided to this study from a group of large fire departments, showed a highly significant difference in *turnout times* between daytime and nighttime hours, a factor not currently addressed in NFPA 1710.

   a. Turnout Times were compared between daytime hours (0600–1800), when crews are presumably at their highest readiness, and nighttime hours (0000–0600), when they are presumably at their lowest readiness.

   b. For both fire and EMS nighttime calls, the mean average turnout times observed fell well above their current NFPA 1710 benchmarks.

      i. For only approximately 21 % of the nighttime fire calls, turnout was completed in the required 80 s or less.

      ii. For only approximately 12 % of the nighttime EMS calls, turnout was completed in the required 60 s or less.

   c. The time required for turnout of 90 % of the nighttime calls was 158 s for fire (just under two times the standard) and 144 s for EMS (slightly more than two and one-third times the standard).

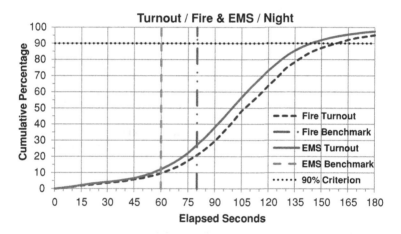

IV. The simulated *turnout times* recorded in the Baseline Turnout Exercise, reported from a diverse group of fire departments, exceeded the benchmarks set in NFPA 1710.

   a. For simulated fire EMS calls, the mean average turnout times observed fell well within their respective current benchmark of 80 s.

      i. For approximately 80 % of the exercise trials using the "wheels rolling" criterion, turnout was completed in the required 80 s or less.

      ii. For approximately 70 % of the exercise trials using the "crosses sill" criterion, turnout was completed in the required 80 s or less.

b. Both percentages of simulated turnouts completed in 80 s or less fall well below the 90 % targeted in the standard.
c. The time actually required and recorded for turnout of 90 % of the calls was 86 s for the "wheels rolling" criterion and 96 s for the "crosses sill" criterion.

Turnout / Fire / Responses & Exercise

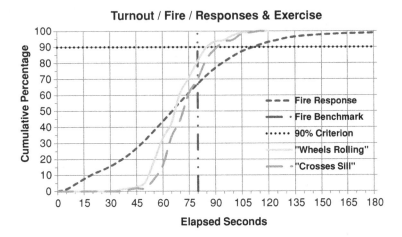

V. The Station Layout Data collected indicate that the average station requires as much as twice the travel distance and time to reach the ERU from common station areas as is provided in the Baseline Turnout Exercise.

a. Foot travel distance and time to sleeping areas is, on the average, significantly greater than travel distance to any other part of the ERF.
b. Foot travel requires 10 s for every 50 feet travelled within the ERF, and stairs more than double that rate.

Station layout (n = 197)	Day room travel (n = 195)	Training room travel (n = 179)	Dining/kitchen travel (n = 195)	Fitness room travel (n = 190)	Sleeping room travel (n = 197)
Minimum	11	6	8	5	13
Median	72	72	69	68	75
Mean	70	71	71	74	84
SD	34	37	30	43	36
Maximum	192	237	155	226	212

Chapter 8
Future Study Questions

- Are there differences in the range of processing complexity for fire versus EMS calls for aid?
- Are there differences in the efficiency of processing algorithms for fire versus EMS calls for aid?
- Does the nature and complexity of calls vary by time of day?
- Are fire versus EMS calls for aid similarly classified and processed as "emergency" responses?
- How much transmission delay is typically introduced between dispatcher initiation and ERF notification of an alarm?
- How does advanced technology impact alarm handling times?

  - CAD to CAD interfaces among PSAPs and response agencies
  - Enhanced mapping data/software
  - Call taker/Dispatcher workflow analysis

- How does the method of alarm notification affect ERU crew turnout times?

  - Automated dispatch messaging/locution systems

- What are the variations in ERU crew turnout times based on the perceived severity of the emergency?

R. Upson and K. A. Notarianni, *Quantitative Evaluation of Fire and EMS Mobilization Times*, SpringerBriefs in Fire, DOI: 10.1007/978-1-4614-4442-8_8, © Fire Protection Research Foundation 2010

Appendix A
Standardized Ideal "Turn Out" Time Drill Layout

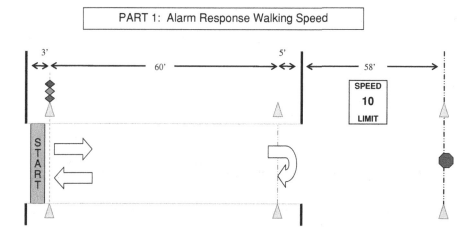

This layout can be reproduced in many typical fire station apparatus bays[1]. This provides a protected indoor environment suitable for measuring firefighter movement speed in a simulated alarm response. The distance from the start and return lines should be measured with a measuring tape or surveyor's wheel as accurately as practical. The start and return lines should be clearly marked. Traffic cones can be used to mark the lines but taped lines on the floor allow for more reliable timing.

---

[1] If the station used does not have room for the full 60′ long course, a 30′ course may be substituted with two laps being completed instead of a single lap. If the alternative layout is used, it must be documented on the data collection form.

R. Upson and K. A. Notarianni, *Quantitative Evaluation of Fire and EMS Mobilization Times*, SpringerBriefs in Fire, DOI: 10.1007/978-1-4614-4442-8,
© Fire Protection Research Foundation 2010

Time over this measured distance will be averaged with data from other departments to estimate an objective movement rate which, in turn, can be used to calculate travel times for various fire station layouts.

Data to track per each trial:

- Trial Identification (date, number)
- Unit Identification (Type, Number, Shift)
- Crew size
- Accurate course completion time for the first and last crew member

A.1 Alarm Response Walking Speed Procedure:

1. The crew moves "quickly and with purpose;" *no* jogging or running; "As you would normally respond to an emergency call in your station."
2. The crew starts on "ready, set, go" command.
3. The crew proceeds to the front end of the bay; crosses over the measured line return; returns and crosses over the starting line to finish. (See footnote 1 for alternate layout)
4. Using a stopwatch, record times for the first and last crew member over the finish line. (If the entire crew crosses the line too closely to reliably time both first and last, record the first time only.)

A.2 Alarm Response Walking Speed Briefing Points

- Move "quickly and with purpose as you would normally respond to an emergency call in your station"
- NO RUNNING OR JOGGING
- Start will be "ready, set, go"—step off quickly

PART 2: Scramble, Don, & Mount

This layout can be reproduced in many typical fire station apparatus bays[2]. It provides a credible simulation of a typical fire station environment with crews in the space immediately adjacent to the apparatus. The close proximity of the crew provides a "best case" scenario which should supply a minimum value for "turnout time."

Split times should be recorded for two possible interpretations of the "en route" criteria:

Data to track per each trial:

- Trial Identification (Date)
- Unit Identification (Type, Number, Shift)
- Crew size
- Time to each "en route" criteria

1. "wheels rolling" measured as soon as the apparatus starts moving
2. "front crosses sill" measured when the front of the apparatus crosses the bay door sill

---

[2] If the station used does not have room for the full 60′ long course, a modified course may be substituted by starting the crew even with the apparatus front bumper on the driver's side and moving to the rear bumper (minimum 30′) before returning and mounting the apparatus instead of the regular layout. If this alternative layout is used, it must be documented on the data collection form.

A.3 Scramble, Don, and Mount Procedure

1. Starting configuration

   - "Le Mans Start"—Crew starts shoulder to shoulder in a line facing away from the apparatus at the back of the bay
   - Crew is dressed in regular station wear
   - PPE is stowed at each crew member's riding position[3]
   - Apparatus is parked with its front 5′ from the line of the door sill
   - MPO & Officer's windows are open
   - Apparatus is otherwise as it would normally be stowed in station

2. On command ("Go!", *no* countdown preparation)

   - *Timer must be positioned to observe "wheels rolling" safely before MPO can release brakes!*

3. Special Tasks

   - Officer retrieves "Run Sheet" from simulated printer
   - MPO opens bay door (manually or with remote)

4. Crew moves promptly to gear; all crew members don core gear at minimum (See footnote 3)

   - Bunker pants
   - Boots
   - Flame retardant hood (*if normally worn*)
   - Bunker coat
   - Helmet (*or headset if that is standard practice*)

5. Crew mounts apparatus

   - MPO may start engine at any time

6. Crew will *not* don SCBA per consensus of technical advisors
7. Crew fastens seatbelts

   - Seatbelt compliance confirmed by apparatus officer
   - MPO shall not release brakes until compliance confirmed

---

[3] EMS crews may omit PPE as per local SOP. Fire suppression crews must don full PPE including either a helmet or radio headset.

8. MPO pulls straight forward promptly (Maximum Speed 10 mph)

- Brakes released; wheels start (**Split 1**)
- Front Bumper crosses doorway sill marked by traffic cones (**Split 2**)
- MPO must *stop* before reaching curb line marked by traffic cones[4]

A.4 Scramble Don, and Mount Briefing Points:

- PPE stowed at each crew member's riding position
- MPO & Officer's windows open
- Apparatus as normally stowed
- Regular station wear *as normally worn*
- "Le Mans Start"

  – Facing away
  – No warning for starting "Go!"

- NO RUNNING OR JOGGING
- Officer retrieves "Run Sheet"
- MPO opens bay door

  – MPO may start engine at any time

- Core gear

  – No SCBA

- Crew fastens seatbelts

  – Compliance confirmed by officer
  – MPO waits for confirmation

- MPO pull forward promptly (10 mph max)

  – *stop* before reaching curb line

---

[4] Where the proximity of public sidewalks and streets limits the distance the apparatus can continue out of the bay, appropriate modifications should be made. Please note any such modifications on the data collection sheet.

Appendix B
Treatment of Extreme Outliers
in the Mobilization Study Historical Data

When all the historical response data from our 14 participating departments was in, we had collected over half a million individual response records. All of these records could not simply be aggregated together due to variations in the ways different departments documented the various time segments critical to our study.

Only data from departments having complete mobilization data—starting with the time the call for assistance was answered, through the time appropriate Emergency Response Units (ERUs) were dispatched, until ERUs were "En route"—was combined. This yielded slightly over 1,53,000 raw data records.

Because of the retrospective nature of the study, it was considered impractical to research individual response records containing what appeared to be idiosyncratic data. To this end, some gross filtering was required on the data prior to analysis. Records where either the Call Processing time or the Turnout Time was recorded as less than or equal to zero were omitted outright as artifacts or documentary errors. It is assumed that these records typically reflect incidents where an ERU came upon an incident without prior dispatch, one incident branched from another, etc. and an incident record was created after the fact.

The second and more difficult to manage concern was for extreme outliers at the upper range of data. In the Call Processing data, for instance, the longest processing time for a fire call was reported at 3,946 s; over 1 h and 5 min! The longest reported processing for an EMS call was not far behind at 3,565 s or just over 59 ½ min. While it is conceivable that some of these extreme outliers represent legitimate Call Processing Times for legitimate emergencies it seems more credible to assume that many of them represent artifacts and non-emergent incidents. The difficult question becomes, which of these extreme outliers can be omitted from the data set to give the most accurate and useful picture of normal Mobilization Times without losing valuable descriptive data?

Figure 1 illustrates graphically the distribution of response data for Call Processing Time for 137,770 fire and EMS responses combined. Only those

R. Upson and K. A. Notarianni, *Quantitative Evaluation of Fire and EMS Mobilization Times*, SpringerBriefs in Fire, DOI: 10.1007/978-1-4614-4442-8,
© Fire Protection Research Foundation 2010

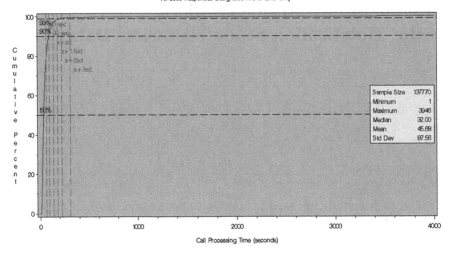

Fig. B.1 CDF combined fire & EMS call processing time (unfiltered)

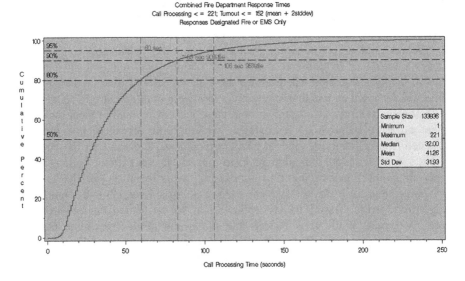

Fig. B.2 CDF combined fire & EMS call processing time

records where Call Processing Time or Turnout Time were less than or equal to zero were omitted in creating this cumulative distribution function (CDF) graph. Starting at the lower left corner at 0 % and 0 s, a blue line representing the cumulative percentage of observed processing responses completed, shown on the

Table B.1 Call Processing Time/Fire (raw)

Fire calls n = 22,564	NFPA 1221 benchmark criteria	Observed compliance	Median Mean Max
Call processing	60 s	79 %	29
	90 %	92 s	56
	90 s	90 %	3946
	99 %	315 s	

Table B.2 Call processing time/fire (filtered)

Fire calls n = 21,954 (610 outliers omitted)	NFPA 1221 benchmark criteria	Observed compliance	Median Mean Max
Call processing	60 s	80 %	29
	90 %	88 s	43
	90 s	91 %	404
	99 %	210 s	

Table B.3 Call processing time/EMS (raw)

EMS calls n = 115,206	NFPA 1221 benchmark criteria	Observed compliance	Median Mean Max
Call processing	60 s	80 %	32
	90 %	84 s	44
	90 s	92 %	3565
	99 %	182 s	

Table B.4 Call processing/EMS (filtered)

EMS calls n = 110,708 (4478 outliers omitted)	NFPA 1221 benchmark criteria	Observed compliance	Median Mean Max
Call processing	60 s	81 %	32
	90 %	79 s	40
	90 s	93 %	154
	99 %	134 s	

vertical axis, is plotted against time, shown on the horizontal axis. The line rises quickly to the median average, 50 % of observed responses, in just 32 s, reaching 90 % of observed responses in just over 90 s, and 99 % of observed responses in a little over 300 s. The last 1 % of the observed responses requires most of the graph for a range of about 3,700 s—over an hour represented in a long "tail" to the right of the otherwise typical CDF of a typical normal distribution.

$$t_{CP} > 0 \,\&\, t_T > 0$$

Table B.5 Turnout time/fire (raw & filtered)

Fire calls n = 22,564	NFPA 1710 benchmark criteria	Observed compliance	Median Mean Max
Turnout	80 s	60 %	71
	90 %	123 s	75
			2629
Fire calls n = 21,954 (610 outliers omitted)	NFPA 1710 benchmark criteria	Observed compliance	Median Mean Max
Turnout	80 s	61 %	71
	90 %	119 s	72
			172

Table B.6 Turnout time/EMS (raw & filtered)

EMS calls n = 115,206	NFPA 1710 benchmark criteria	Observed compliance	Median Mean Max
Turnout	60 s	54 %	58
	90 %	109 s	63
			3112
EMS calls n = 110,708 (4478 outliers omitted)	NFPA 1710 benchmark criteria	Observed compliance	Median Mean Max
Turnout	60 s	55 %	57
	90 %	103 s	60
			147

The red vertical lines overlaid on the graph indicate the mean average times of observed responses, 46 s, plus various multiples of the standard deviation of 88 s (mean plus 1, 1 ½, 2, and 3 standard deviations) as various candidates for initial cutoff filtering of extreme outliers.

The mean average plus 2 standard deviations, 221 s in this example, was chosen as an initial cutoff filter for comparison. In a normal distribution, this would omit 2.28 % of the data. A CDF of the same data omitting responses where either the Call Processing Time or Turnout Time exceed the mean plus 2 standard deviations of the raw data is shown in Fig. 2. The filtered data produces a more visually informative graph, retains its original median value, and arguably represents a more credible range of processing times for actual emergencies.

$$t_{CP} > 0 \, \& \, t_T > 0$$

$$t_{CP_{filtered}} \leq \left(\bar{x}_{CP_{raw}} + 2S_{CP_{raw}}\right) \& t_{T_{filtered}} \leq \left(\bar{x}_{T_{raw}} + 2S_{T_{raw}}\right)$$

B.3 Raw versus Filtered Data

Analyzing 22,564 raw fire response records yields the results shown in Table 1. In this analysis, 79 % of the observed fire responses meet the 60 s NFPA 1221 benchmark for call processing and 90 % meet the 90 s benchmark. The median average, representing 50 % of all calls, is 29 s.

Applying a mean plus 2 standard deviation filter in Table 2 omits 610 response records or about 2.7 % of the total. The changes in the calculated compliance times and percentages only shift by 1 s and 1 % respectively with no change in the median average.

Performing a similar analysis on 115,206 raw EMS response records yields the results shown in Table 3. In this analysis, 80 % of the observed EMS responses meet the 60 s NFPA 1221 benchmark for call processing and 92 % meet the 90 s benchmark. The median average, representing 50 % of all calls, is 32 s.

Applying a mean plus 2 standard deviation filter in omits 4478 response records or about 3.9 % of the total. Once again, the changes in the calculated compliance times and percentages only shift by 1 s and 1 % respectively with no change in the median average as shown in Table 4.

The same pattern can be observed with Turnout Time for raw versus filtered fire response calls in Table 5 and also with Turnout Time for raw versus filtered fire response calls in Table 6.

As the only other measure taken from the Historical Response Data, Mobilization Time, is created by summing Call Processing Time and Turnout Time, it shows the same stability when comparing raw and filtered statistics.

Because the filtering of data only presumably removes some artifacts and atypical outliers at the expense of also presumably removing some valid information, there should be a clear benefit to justify doing so. No such benefit was found with respect to this study with the exception on using filtered data for the creation of CDF plots in order to magnify the critical areas of the plot and minimize the data "tail" created by extreme outliers.

About The Authors

Bob Upson

Robert Upson is a graduate student at the Worcester Polytechnic Institute (WPI), where he is pursuing a Master of Science degree in Fire Protection Engineering. Upson is currently developing his thesis project on a comprehensive analysis of fire service turnout time. He successfully authored and managed an Assistance to Firefighters Grant for $73,000 in 2007 for his fire department's public education program. Prior to entering the fire service, Upson worked as a database manager and research assistant  for a large university research project. Upson recently completed data collection for a pilot study on fire service turnout times in Montgomery County, MD, as an adjunct to the ongoing "Multi-Phase Study on Firefighter Safety and Deployment of Resources." Upson holds a B.A. in Psychology from the University of Connecticut, a B.S. in Fire Science Technology from Charter Oak State College, New Britain, CT, and numerous fire service professional certifications. He is currently a deputy fire marshal and career firefighter with over 25 years on the job.

R. Upson and K. A. Notarianni, *Quantitative Evaluation of Fire and EMS Mobilization Times*, SpringerBriefs in Fire, DOI: 10.1007/978-1-4614-4442-8,

Kathy Notarianni, PhD., PE

Kathy A. Notarianni is the Head of the Department of Fire Protection Engineering at Worcester Polytechnic Institute (WPI). Notarianni works with the university's fire protection engineering faculty to plan for the future of graduate studies and research in fire protection engineering, which incorporates elements of civil, structural, electrical, and chemical engineering to make structures, vehicles, clothing, and people safer from fire. She strives to build strong networks with agencies, laboratories, universities, and companies that have a common  interest in fire protection engineering education and research. Prior to joining WPI, Notarianni managed a group of scientists and engineers in a technical program of integrated performance assessment and risk at the National Institute of Standards and Technology (NIST) in Gaithersburg, MD. The program serves to produce tools to quantify fire events for fire hazard and risk assessment; fire fighting operations and training; fire investigations and performance evaluations of fire protection systems in buildings; and transportation networks and vehicles in response to fire. Notarianni is well published, having authored or co-authored more than 30 publications, including chapters in two books, the *Handbook of Fire Protection Engineering* and *Improving Regulations*. She has been recognized by the Society of Fire Protection Engineers (SFPE) and U.S. Department of Commerce with awards for leadership and performance. Notarianni holds a B.S. in chemical engineering and a M.S. in fire protection engineering, both from WPI. She later earned a PhD in engineering and public policy from Carnegie Mellon University, where she did her doctoral dissertation on "The Role of Uncertainty in Improving Regulation: A Case Study in Fire Protection." She has been awarded over $5 million in research grants from multiple governmental sponsors such as NASA, NIH, and the U.S. Navy. She is a fellow in the SFPE. Notarianni is currently one of three Principle Investigators on a $3 million study of Firefighter Safety and Resource Deployment sponsored by the U.S. Department of Homeland Security.

Works Cited

Averill JD, Moore-Merrell L, Notarianni KA, Santos R, Barowy A. Multi-phase study on firefighter safety and deployment study year 1 final report. FireReporting.org. 19 Sept 2008. http://www.firereporting.org/pdfs/2008_Year_1_Final_Report.pdf. Accessed March 2010

Boyce, K.E., Shields, T.J., Silcock, G.W.H.: Toward the characterization of building occupancies fore fire safety engineering: capabilities of disabled people moving horizontally and on an incline. Fire Technol 35(1), 51–67 (1999)

Bryan JL (2002) Behavioral response to fire and smoke. Chap. 12. In: DiNenno PJ (ed) The SFPE handbook of fire protection engineering, Quincy, NFPA

Bryan JL (2008) Human behavior and fire. Vol. 1, Chap. 4–1 In: Cote AE (ed) Fire protection handbook, 4-3-4-47. Quincy, NFPA

CPSE/Deccan International (2007) Turnout investigation. Internal report to NFPA 1710 Technical Committee, Center for Public Safety Excellence, unpublished

Gill C (2009) Unpublished correspondence, 13 Feb 2009

Gill C (2007) IRMP year III—Turnout times. Royal berkshire fire and rescue service, 23 April 2007. http://library.rbfrs.co.uk/public_agendas/osrqj100.doc Accessed March 2010

Klaene BJ, Sanders RE (2008) NFPA structural firefighting strategy and tactics. 2nd. Sudbury, Jones and Bartlett Publishers

Microsoft Corporation (2006) Excel 2007. Part of microsoft office small business edition 2007

NFPA 1221 (2009) NFPA 1221, standard for the installation, maintenance, and use of emergency services communications systems. 2010. Quincy, NFPA

NFPA 1710 (2009) NFPA 1710, standard for the organization and deployment of fire suppression operations, emergency medical operations, and special operations to the public by career fire departments. 2010. Quincy, NFPA

NFPA 1710 ROC. (2009) 2009 Annual Revision Cycle Report on Comments. Quincy NFPA, 1710-5–1710-10

Office of Strategic Planning and Information Systems of the Greensboro (NC) Fire Department (2007) Guilford Metro 911 Emergency Communications Center. MCT Enroute 800MHz Timing Study. internal study, Office of Strategic Planning and Information Systems, Greensboro (NC) Fire Department, NC

SAS Institute Inc (2003) SAS. Cary, SAS Institute Inc.

The IAFF/IAFC Wellness-Fitness Task Force (2010) IAFF: wellness-fitness initiative. http://www.iaff.org/hs/CPAT/cpat_index.html. Accessed March 2010

U.S. Fire Administration (2008) USFA fire departments. http://www.usfa.dhs.gov/statistics/departments/index.shtm. Accessed March 2010

Upson RPS (2009) Turnout validation study. Independent study project, fire protection engineering, Worcester Polytechnic Institute, Worcester, unpublished